EDEXCEL
FURTHER MATHS

FURTHER MECHANICS 1
For AS and A Level

Series Editor
David Baker

Authors
Brian Jefferson, Mike Heylings
David Bowles, Eddie Mullan, Garry Wiseman, John Rayneau, Katie Wood, Rob Wagner

OXFORD
UNIVERSITY PRESS

OXFORD
UNIVERSITY PRESS

Great Clarendon Street, Oxford, OX2 6DP, United Kingdom

Oxford University Press is a department of the University of Oxford.

It furthers the University's objective of excellence in research, scholarship, and education by publishing worldwide. Oxford is a registered trade mark of Oxford University Press in the UK and in certain other countries.

British Library Cataloguing in Publication Data
Data available

978 0 19 841529 9

10 9 8 7 6 5 4 3 2 1

Paper used in the production of this book is a natural, recyclable product made from wood grown in sustainable forests.
The manufacturing process conforms to the environmental regulations of the country of origin.

Printed in Great Britain by CPI Group (UK) Ltd, Croydon, CR0 4YY

Acknowledgements
Authors
Brian Jefferson, Mike Heylings
David Bowles, Eddie Mullan, Garry Wiseman, John Rayneau, Katie Wood, Rob Wagner

Editorial team
Dom Holdsworth, Ian Knowles, Matteo Orsini Jones, Rosie Day

With thanks also to Geoff Wake, Matt Woodford, Katherine Bird for their contribution.

Index compiled by Marian Preston, Preston Indexing.

Although we have made every effort to trace and contact all copyright holders before publication, this has not been possible in all cases. If notified, the publisher will rectify any errors or omissions at the earliest opportunity.

p1 Dolomites-image/istockphoto, **p22** Andyd/iStockphoto, **p22** Steve Mann/Dreamstime, **p27** FatCamera/iStockphoto, **p58** selensergen/iStockphoto, **p58** Tatiana Shepeleva/Shutterstock.

Contents

Chapter 1: Forces and energy

Introduction.. 1

1.1 Work, energy and power 2

1.2 Hooke's Law 12

Summary and review 19

Exploration.. 22

Assessment 23

Chapter 2: Momentum and collisions

Introduction.. 27

2.1 Conservation of momentum 28

2.2 Collisions 34

2.3 Impulses 40

2.4 Momentum and impulse in two dimensions 46

Summary and review 55

Exploration.. 58

Assessment 59

Mathematical formulae.. **64**

Mathematical formulae – to learn... **72**

Mathematical notation.. **79**

Answers **84**

Index... **98**

About this book

This book has been specifically created for those studying the Edexcel 2017 Further Mathematics AS and A Level. It's been written by a team of experienced authors and teachers, and it's packed with questions, explanation and extra features to help you get the most out of your course.

Every section starts by covering the basic **Fluency and skills** (A01).

Key points highlight important concepts, and make the information easier to digest.

Worked examples provide a model answer and commentary to realistic practice questions.

There is a Fluency and skills exercise for each section, to practise the skills before moving on to the Reasoning and problem-solving section.

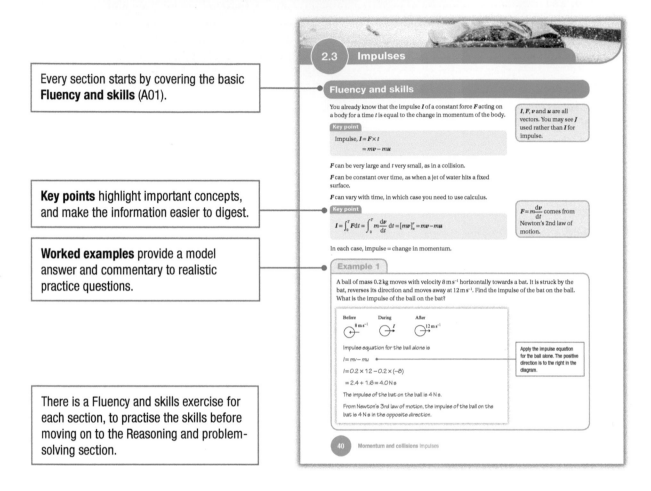

On the chapter **Introduction page**, the Orientation box explains what you should already know, what you will learn, and what this leads to.

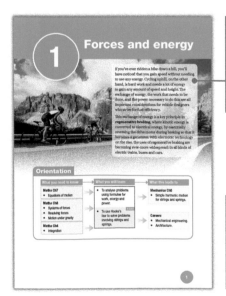

At the end of every chapter, an **Exploration page** gives you an opportunity to explore the subject beyond the specification.

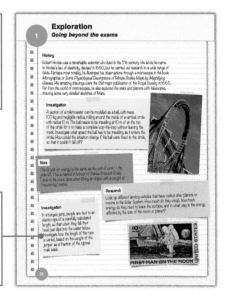

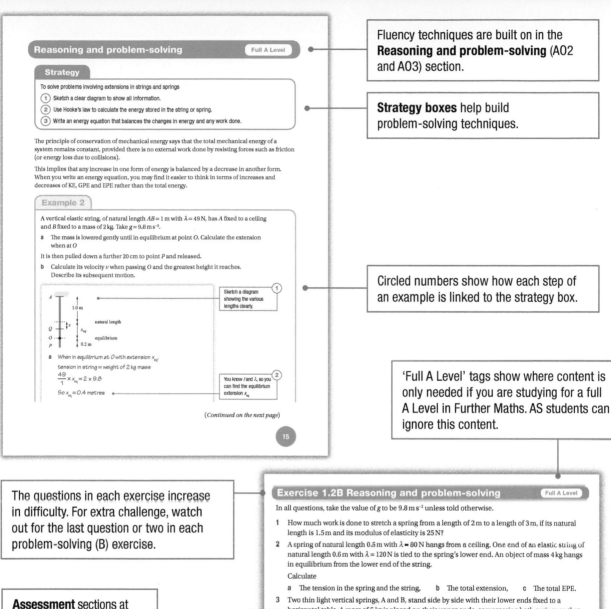

Fluency techniques are built on in the **Reasoning and problem-solving** (AO2 and AO3) section.

Strategy boxes help build problem-solving techniques.

Circled numbers show how each step of an example is linked to the strategy box.

'Full A Level' tags show where content is only needed if you are studying for a full A Level in Further Maths. AS students can ignore this content.

The questions in each exercise increase in difficulty. For extra challenge, watch out for the last question or two in each problem-solving (B) exercise.

Assessment sections at the end of each chapter test everything covered within that chapter.

Reasoning and problem-solving — Full A Level

Strategy

To solve problems involving extensions in strings and springs

1. Sketch a clear diagram to show all information.
2. Use Hooke's law to calculate the energy stored in the string or spring.
3. Write an energy equation that balances the changes in energy and any work done.

The principle of conservation of mechanical energy says that the total mechanical energy of a system remains constant, provided there is no external work done by resisting forces such as friction (or energy loss due to collisions).

This implies that any increase in one form of energy is balanced by a decrease in another form. When you write an energy equation, you may find it easier to think in terms of increases and decreases of KE, GPE and EPE rather than the total energy.

Example 2

A vertical elastic string, of natural length $AB = 1\,\text{m}$ with $\lambda = 49\,\text{N}$, has A fixed to a ceiling and B fixed to a mass of 2 kg. Take $g = 9.8\,\text{m s}^{-2}$.

a The mass is lowered gently until in equilibrium at point O. Calculate the extension when at O

It is then pulled down a further 20 cm to point P and released.

b Calculate its velocity v when passing O and the greatest height it reaches. Describe its subsequent motion.

① Sketch a diagram showing the various lengths clearly.

a When in equilibrium at O with extension x_{eq},
tension in string = weight of 2 kg mass
$$\frac{49}{1} \times x_{eq} = 2 \times 9.8$$
So $x_{eq} = 0.4$ metres

② You know l and λ, so you can find the equilibrium extension x_{eq}

(Continued on the next page)

15

Exercise 1.2B Reasoning and problem-solving — Full A Level

In all questions, take the value of g to be $9.8\,\text{m s}^{-2}$ unless told otherwise.

1 How much work is done to stretch a spring from a length of 2 m to a length of 3 m, if its natural length is 1.5 m and its modulus of elasticity is 25 N?

2 A spring of natural length 0.5 m with $\lambda = 80\,\text{N}$ hangs from a ceiling. One end of an elastic string of natural length 0.6 m with $\lambda = 120\,\text{N}$ is tied to the spring's lower end. An object of mass 4 kg hangs in equilibrium from the lower end of the string.

Calculate

a The tension in the spring and the string,   b The total extension,   c The total EPE.

3 Two thin light vertical springs, A and B, stand side by side with their lower ends fixed to a horizontal table. A mass of 5 kg is placed on their upper ends, compressing both springs so that the mass is in equilibrium. If the natural length and modulus of elasticity of A are 0.8 m and 80 N, and those of B are 1.5 m and 60 N, find

a The compressed length of the springs,

b The tension in each spring,

c The total EPE in the springs.

---

2 Assessment

1 An object of mass 7 kg, travelling at a speed of $4\,\text{m s}^{-1}$, is acted on by a constant force in its direction of travel which increases its speed to $10\,\text{m s}^{-1}$. Calculate

a The impulse exerted on the object,   **[2 marks]**

b The force involved if the process took 0.35 seconds.   **[2]**

2 A particle of mass 2 kg is travelling in a straight line at $8\,\text{m s}^{-1}$. A variable braking force acting along the same line of travel is applied so that, after t s, the magnitude of the force is $2t$ N. Calculate the time taken for the particle to come to rest.   **[3]**

3 A particle of mass 5 kg is travelling with a velocity of $(4\mathbf{i} + \mathbf{j})\,\text{m s}^{-1}$ when it is subjected to an impulse of $(2\mathbf{i} - 7\mathbf{j})\,\text{N s}$. Calculate the new velocity of the particle.   **[3]**

4 A particle of mass 3 kg has velocity $(2\mathbf{i} - 3\mathbf{j})\,\text{m s}^{-1}$. It is acted on by a constant force of $(-\mathbf{i} + 2\mathbf{j})\,\text{N}$, which changes its velocity to $0.5\mathbf{i}\,\text{m s}^{-1}$. For how long does the force act?   **[4]**

10 A sledgehammer of mass 6 kg, travelling at $20\,\text{m s}^{-1}$, strikes the top of a post of mass 2 kg, which rests in soft ground, and maintains contact with the post.

a Calculate the common speed of the hammer and post immediately after impact.   **[2]**

b The post is brought to rest in 0.02 s by the action of a resistive force R from the ground. By modelling R as constant, find its magnitude.   **[2]**

c If, in fact, the force is given by $R = k(1 + 2t)\,\text{N}$, where t s is the time from the moment of impact, find the value of the constant k.   **[3]**

d In what ways would the situation be different if the sledgehammer were to rebound on impact?   **[2]**

11 An object of mass 3 kg has velocity $(3\mathbf{i} + 2\mathbf{j})\,\text{m s}^{-1}$. It collides with another object, which has a mass of 2 kg and a velocity of $(\mathbf{i} - \mathbf{j})\,\text{m s}^{-1}$. After the impact, the first object has a velocity of $(2\mathbf{i} + \mathbf{j})\,\text{m s}^{-1}$. Calculate the velocity of the second object.   **[2]**

12 Particles A and B have masses of 3 kg and 2 kg respectively. They are connected by a light inextensible string. The particles lie at rest on a smooth horizontal surface. The coefficient...

Full A Level

Forces and energy

1

If you've ever ridden a bike down a hill, you'll have noticed that you gain speed without needing to use any energy. Cycling uphill, on the other hand, is hard work and needs a lot of energy to gain any amount of speed and height. The exchange of energy, the work that needs to be done, and the power necessary to do this are all important considerations for vehicle designers who strive for fuel efficiency.

This exchange of energy is a key principle in **regenerative braking**, where kinetic energy is converted to electrical energy, by essentially reversing the drive motor during braking so that it becomes a generator. With electronic technology on the rise, the uses of regenerative braking are becoming ever-more widespread: in all kinds of electric trains, buses and cars.

Orientation

What you need to know	What you will learn	What this leads to
Maths Ch7 • Equations of motion	• To analyse problems using formulae for work, energy and power.	**Mechanics Ch5** • Simple harmonic motion for strings and springs.
Maths Ch8 • Systems of forces • Resolving forces • Motion under gravity	• To use Hooke's law to solve problems involving strings and springs. _A Level_	**Careers** • Mechanical engineering. • Architecture.
Maths Ch4 • Integration		

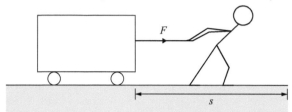

Fluency and skills

If you want to move an object, you have to push or pull it and so do some work. How much work you do depends on the force F you exert and the distance s you move *in the direction of the force*.

Key point

Work done by a constant force = force × distance
$$= F \times s$$
when F and s are in the same direction.

The units of work are **joules** (J), assuming force is in N and distance in m.

When the force F is not constant, but varies over the distance s moved in the direction of F, the work done by F over a small distance δs is given by $F \times \delta s$

Key point

Summing over the total distance s gives:

the total work done by a variable force = $\displaystyle\int_0^s F \, ds$

If you drag an object horizontally along a rough floor for a distance s, four forces are acting on the object.

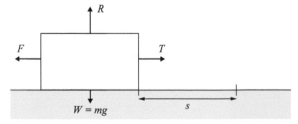

> Joules (J) are the same as N m.

> The letter F is often used for any general force, and also for a frictional force.
>
> The letter R is often used for a reaction force and also for a resisting force.

> Use the word *by* when the force does work and the word *against* when work is done against a resisting force such as friction.

The object is dragged by a force T (the tension in a rope or the tractive force of an engine) against a frictional force F. These forces act in the same straight line as the motion of the object, so the work done *by* the dragging force over distance s is $T \times s$, and the work done *against* the frictional force is $F \times s$ (if T and F are constant).

Power is the rate at which work is done. If an engine is set to work for a period of time, then the power of the engine is equal to $\dfrac{\text{work done}}{\text{time taken}}$.

The units of power are **watts** (W) when work is in J and time in s.

If a constant force F N pulls an object in its direction at a constant speed v m s^{-1}, the object moves a distance of v metres every second. For example, a train where an engine is pulling coaches, with force F, will move at a steady speed v

The work done every second is $F \times v$ joules, so the power of the engine is $F \times v$ watts.

> This formula gives the average power over the given time. In a question involving force or power, you should assume the average value is being discussed unless stated otherwise.

Key point

> The power of a constant force F moving at a steady speed v equals $F \times v$, where power is in W, force in N and speed in m s^{-1}.
>
> If F or v are variable, $F \times v$ gives the power at that particular instant.

There are many kinds of energy, but mechanics is concerned with two kinds: **kinetic energy** (due to a body's motion) and **potential energy** (due to a body's position).

When you apply a force F to a body of mass m which is initially at rest, it moves with acceleration a. After travelling a distance s, it has velocity v

Newton's second law gives $\qquad\qquad\qquad\qquad F = ma$

The kinematic equation $v^2 = u^2 + 2as$ gives $\qquad v^2 = 2as$ because $u = 0$

So the work done by the force F is $\qquad\qquad F \times s = ma \times \dfrac{v^2}{2a} = \dfrac{1}{2}mv^2$

The KE of the body due to the motion is measured by the amount of work done by the force F

Key point

> The **kinetic energy** (KE) of a body with mass m and velocity v is given by $\text{KE} = \dfrac{1}{2}mv^2$

When you release a mass m from height h, it falls freely under gravity while being acted on by a force F

Newton's second law gives $F = ma$ where a is the acceleration due to gravity, g

So the force is its weight, $F = mg$ and the work done by F is $F \times h = mgh$

Before the mass is released, the force of gravity has potential to do work on the mass. So, the energy it possessed due to its height before being released is called its gravitational potential energy.

Key point

> The **gravitational potential energy** (GPE) of a body with mass m at height h is equal to mgh

The **principle of conservation of mechanical energy** says that the total mechanical energy (the sum of KE and GPE) of a system remains constant, provided there is no external work done on energy lost due to any impacts.

For example, to keep the total energy constant, the loss of GPE when an object falls equals its gain of KE provided no energy is lost to air resistance. However, if energy is lost, you would have, for this example, the energy equation:

Loss of GPE = Gain in KE + Work done against air resistance

More generally, a body may gain energy due to tractive forces as well as losing energy due to friction or other resistive forces. So a general equation for mechanical energy is:

Work done by tractive forces = Gain in KE + Gain in GPE + Work done against resistances

Example 1

A crate of mass 80 kg is pulled at a constant speed of $0.2\,\mathrm{m\,s^{-1}}$ for 3 metres across a rough horizontal floor by a rope parallel to the floor. There is a frictional force F of $32g\,$N.

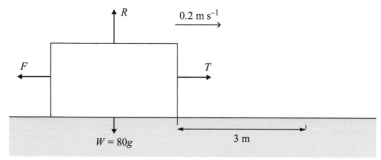

Calculate, in terms of g,

a The work done by each force acting on the crate,

b The kinetic energy of the crate,

c The power of the tension T in the rope.

There is no horizontal or vertical acceleration.

Reaction R = weight W = $80g\,$N ●————

Tension T = friction F = $32g\,$N ●——

> Resolve vertically and horizontally.

a There is no vertical distance moved,
so work done by R and W is zero.

The horizontal distance moved = 3 m

So, work done by $T = 32g \times 3 = 96g\,$J ●————

> Calculate the work done to move the crate against friction.

Friction opposes the tension T, so the work done by T is used to overcome friction.

Work done against $F = 32g \times 3 = 96g\,$J.

b Kinetic energy of the crate $= \frac{1}{2}mv^2 = \frac{1}{2} \times 80 \times 0.2^2 = 1.6\,$J

c Rate of working = Power = $T \times v = 32g \times 0.2 = 6.4g\,$W

Example 2

A car of mass 800 kg starts from rest at the foot of a slope. It climbs the slope, travelling 50 m in 5 s to reach a speed of 20 m s⁻¹ as it rises a vertical height of 4 m.

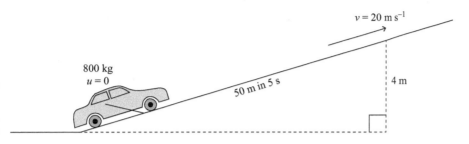

Stating your assumptions and taking $g = 10 \text{ m s}^{-2}$, calculate

a The gain in the car's GPE and KE,

b The tractive force and power exerted by its engine.

Assume that negligible energy is lost to friction or air resistance.

a Gain in GPE $= mgh = 800 \times 10 \times 4 = 32\,000$ J

Gain in KE $= \dfrac{1}{2}mv^2 = \dfrac{1}{2} \times 800 \times 20^2 = 160\,000$ J

b Let the tractive force of the engine be T newtons.

Work done by the engine $= T \times s = T \times 50$ J

The energy equation is

Gain in GPE + Gain in KE = Work done by engine

$32\,000 + 160\,000 = T \times 50$

So, tractive force $T = \dfrac{192\,000}{50} = 3840$ N

Power of the engine $= \dfrac{\text{Work done}}{\text{Time taken}} = \dfrac{192\,000}{5}$

$= 38\,400$ W $= 38.4$ kW

> Work done by the external force is equal to the overall increase in energy.
> Here, work is done against gravity and to increase the speed of the car.

If you drag an object along a floor for a distance s, and the constant tractive force T is at an angle θ to the distance, you can calculate the work done by T in two ways:

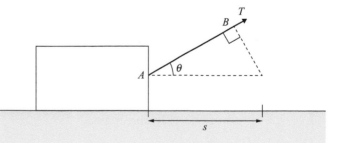

- *Either* use the distance in the direction of $T = AB = s\cos\theta$, giving
  Work done $= T \times s\cos\theta = Ts\cos\theta$
- *Or* use the component of T in the direction of motion, giving
  Work done $= T\cos\theta \times s = Ts\cos\theta$

Work done by a constant force F is given by $Fs\cos\theta$ when the angle between F and s is θ

Exercise 1.1A Fluency and skills

In all questions, take the value of g to be $9.8\,\mathrm{m\,s^{-2}}$ unless told otherwise.

1 Calculate the work done and power required when

    a A crate is moved at a steady pace for 6 m, over a period of 4 s, and in the horizontal direction by a horizontal rope under a tension of 120 N,

    b A spring balance that reads 20 N steadily lifts a suitcase by 3 m in 1.5 seconds.

2 A variable horizontal force F pulls an object along a horizontal table through a distance s m. If $s = 4$, calculate the work done by the force if

    a $F = 8 - 2s$ **b** $F = 16 - s^2$ **c** $F = s^2 - s + 2$

3 A horizontal rope pulls a trolley 6 m along a horizontal floor. A graph is drawn to show how the tension T in the rope varies. Calculate the work done by the tension in each case.

    a 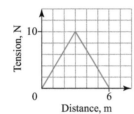   **b**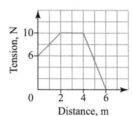

4 A winch uses a rope to lift three boxes A, B and C, at constant speeds over these distances and times.

Box	A	B	C
Mass, kg	12	4	0.5
Distance, m	3	0.5	0.2
Time, s	4	4	2.0

For each box, find, in terms of g,

    a The tension T in the rope,

    b The work done by the winch,

    c The constant power the winch exerts,

    d The gain in GPE of the box.

5 **a** A lorry of mass 2 tonnes accelerates from rest to a speed of $10\,\mathrm{m\,s^{-1}}$ over 400 metres on a horizontal road with negligible resistance. Calculate

        i The gain in its KE,

        ii The work done by the engine,

        iii The engine's tractive force.

    b If the lorry had experienced a constant resistance of 50 N, how much work would the engine now do and what would be the tractive force?

6   A train of mass 50 tonnes increases its speed from $10\,\mathrm{m\,s^{-1}}$ to $20\,\mathrm{m\,s^{-1}}$ over a horizontal distance of 500 metres against a resistance of $2000\,\mathrm{N}$ due to wind and friction.

   a   Calculate the increase in the train's KE.

   b   Calculate the work done by the engine and the engine's tractive force.

   c   What is the initial and final power of the engine?

7   a   A brick of mass $1.2\,\mathrm{kg}$ falls from rest from the top of a building 30 metres tall.

      i   What is the GPE lost and the KE gained just before hitting the ground?

      ii   Calculate the brick's speed on impact.

   b   If the brick is thrown down from the top of the building with an initial speed of $14\,\mathrm{m\,s^{-1}}$, what is its total KE and its speed just before impact?

   c   If the brick is thrown upwards from the top of the building with a speed of $14\,\mathrm{m\,s^{-1}}$, what is its speed on impact with the ground?

   d   If the brick experiences a constant resistance of $3\,\mathrm{N}$ in falling from rest, what is its velocity when striking the ground?

8   A car of mass 1.2 tonnes travels up a slope at 30° to the horizontal at a constant speed of $25\,\mathrm{m\,s^{-1}}$ for 300 metres.

   a   If there is no resistance to motion, calculate

      i   The increase in the car's energy,

      ii   The tractive force of its engine,

      iii Its power output.

   b   If the car experiences a constant resistance to motion of $3000\,\mathrm{N}$, calculate the tractive force and power of its engine.

Reasoning and problem-solving

Strategy

To solve problems involving work, energy and power

(1) Draw a clearly labelled diagram to show the information you are given.

(2) Write an energy equation, ensuring you have the gains and losses balanced correctly.

Example 3

A car of mass 1 tonne travels from rest up a slope at 30° to the horizontal with a constant acceleration against a constant resistance R of 400 N. On reaching the top of the slope, it has a speed of $10\,\mathrm{m\,s^{-1}}$ and its engine is working at a rate of 58 kW. Calculate the length of the slope.

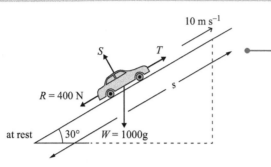

① Draw a clear diagram showing all the information you have been given.

You may also find it helpful to list the values you know and what you need to find, to help you choose the equations to use.

The acceleration is constant, so the tractive force T of the engine is constant.

Power $= Tv$

$58000 = T \times 10 \Rightarrow T = 5800\,\mathrm{N}$

Let s = the length of the slope.

Using the right-angled triangle, the vertical height gained is $\sin 30°\, s$

Use the trigonometric ratio $\sin\theta = \dfrac{\text{opposite}}{\text{hypotenuse}}$

There are now two methods.

<u>Using work and energy</u>

KE gained $= \dfrac{1}{2}mv^2 = \dfrac{1}{2} \times 1000 \times 10^2 = 50000\,\mathrm{J}$

GPE gained $= mgh = 1000 \times 9.8 \times \sin 30°\, s = 4900s$

Work done by engine $= 5800 \times s\,\mathrm{J}$

Work done against $R = 400 \times s\,\mathrm{J}$

Calculate the change in energy.

The car gains both KE and GPE.

Energy equation is

Work done by engine = KE gained + GPE gained + Work done against R

$5800s = 50000 + 4900s + 400s$

$500s = 50000$

Length of slope, $s = 100$ metres

② The work done by the engine increases the speed and height of the car and overcomes resistance.

<u>Using Newton's 2nd law</u>

The component of weight down the slope is $mg\sin 30°$

Equation of motion along the slope is

$T - R - mg\sin 30° = ma$

$5800 - 400 - 1000 \times 9.8 \times 0.5 = 1000 \times a$

$500 = 1000a$

Acceleration $a = 0.5\,\mathrm{m\,s^{-2}}$

Using $v^2 = u^2 + 2as,$

$10^2 = 0 + 2 \times 0.5 \times s$

Length of slope, $s = 100$ metres

Use the constant acceleration formula.

In all questions, take the value of g to be $9.8\,\text{m s}^{-2}$ unless told otherwise.

1 A train of mass 100 tonnes accelerates steadily from rest over 450 metres to reach a speed of $15\,\text{m s}^{-1}$. The resistance to motion is negligible.

    a Use an energy equation to calculate the maximum power exerted by its engine.

    b Calculate the acceleration of the train and use Newton's 2nd law to check your answer.

    c If the resistance to motion were not negligible, how would taking resistance into account affect your answers to parts **a** and **b**? Explain your reasoning.

2 The same train as in Question **1** reaches the same speed from rest over the same distance but resistance to its motion is $4000\,\text{N}$. Use an energy equation to calculate the maximum power now exerted by its engine. Check your answer using Newton's 2nd law.

3 Particles P of mass $2\,\text{kg}$ and Q of mass $5\,\text{kg}$ are connected by a light inextensible string passing over a light smooth pulley. They are initially at rest at the same level with the string taut and then released.

    a Use an energy equation to calculate the speed of the particles when they are 3 metres apart.

    b State two limitations to the model used to solve this problem.

4 A cyclist and her bike have a total mass of $80\,\text{kg}$. The cyclist free-wheels $200\,\text{m}$ downhill against a constant resistance while dropping a vertical distance of $8\,\text{m}$. Her speed increases from $3\,\text{m s}^{-1}$ to $10\,\text{m s}^{-1}$. Calculate the gain in kinetic energy and hence calculate the resistance to the motion.

5 The maximum power of the engine of a 5 tonne coach is $40\,\text{kW}$. When the resistance to motion is $1500\,\text{N}$, calculate the maximum speed it can reach on the level.

6 A fountain uses a pump to raise 3000 litres of water through 4 metres every minute and expels it at a speed of $8\,\text{m s}^{-1}$. Find the power of the pump. The density of water is 1 kg per litre.

7 Water is discharged at $6\,\text{m s}^{-1}$ through a circular nozzle of radius $4\,\text{cm}$, having been raised by $2\,\text{m}$. Find the power of the pump to the nearest watt. The density of water is 1 kg per litre.

8 A train of mass $1200\,\text{kg}$ with velocity $v\,\text{m s}^{-1}$ is resisted by a force of $(600 + v^2)\,\text{N}$. The power of its engine is $1.5v\,\text{kW}$. When on level ground, calculate

    a Its maximum speed,

    b Its acceleration when travelling at a speed of $10\,\text{m s}^{-1}$.

9 The total mass of an engine and its train is $450\,000\,\text{kg}$. The resistance to their motion is $\dfrac{v^2}{4}\,\text{N}$ per $1000\,\text{kg}$ at speed $v\,\text{m s}^{-1}$. If the maximum power of the engine is $1125\,\text{kW}$, calculate its greatest speed on horizontal ground.

10 A cyclist on a level road has a power of $75\,\text{W}$ when travelling at constant speed $v\,\text{m s}^{-1}$ against resistance $R = kv^2\,\text{N}$. Calculate the value of the constant k if the speed is $24\,\text{km h}^{-1}$.

11 The resistance to the motion of a car is directly proportional to its speed. A car of mass $1500\,\text{kg}$ has a maximum speed of $45\,\text{m s}^{-1}$ on level ground when its engine works at a rate of $8\,\text{kW}$. Calculate its acceleration when it moves at $20\,\text{m s}^{-1}$ on the level if the power of its engine is $6\,\text{kW}$.

12 An object of mass 8 kg, powered by an engine, is dragged along a rough horizontal floor from rest for a distance s metres by a variable force T where $T = 36 - s^2$ newtons.

   a   Find the object's speed and the power of its engine after moving 4 metres, given that the coefficient of friction between the object and the floor, μ, $= 0.25$

   b   For what range of values of s is this model for T valid? Suggest a different model that would be valid for values of s up to 10 metres.

13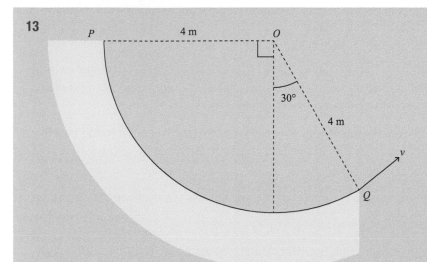

   A skateboarder of mass 60 kg goes from rest at P down a ramp shaped as an arc of a circle, centre O, radius 4 m. Calculate his speed v on leaving the ramp at Q when there is

   a   No resistance to motion,

   b   A constant resistance to motion of 48 N.

   c   i   How would your answer for v be affected if friction were taken into account? Give your reasoning.

       ii Is it reasonable to model the resistance to motion as constant? Describe an alternative to improve this modelling of air resistance.

14 For the coach in question **5**, what is its maximum speed up a slope that makes an angle of $\sin^{-1}\dfrac{1}{98}$ to the horizontal? The resistance to motion is the same as in question **5**

15 A car of mass 900 kg pulls a trailer of mass 200 kg and the resistances to their motion are constant at 200 N and 80 N respectively. When travelling on the level at a maximum speed of 28 m s^{-1}, the car engine is working at maximum power. If, instead, it is travelling with a speed of 8 m s^{-1} at full power uphill at an angle of $\arcsin\dfrac{1}{40}$ to the horizontal, calculate

   a   The acceleration,

   b   The tension in the tow-bar.

16 A smooth ring of mass m is threaded on a fixed vertical circular hoop with centre O and radius r. It is projected with velocity u from its lowest point.

    a Show that, if it just reaches the highest point Q, then $u^2 = 4gr$

    b If it passes point P where acute angle $QOP = \theta$ with velocity $\frac{1}{2}u$, prove that $\cos\theta = \dfrac{3u^2}{8gr} - 1$

    and that $\dfrac{8gr}{3} < u^2 < \dfrac{16gr}{3}$

17 For the cyclist in question **10**, calculate the steepest angle she can climb at $12\,\text{km}\,\text{h}^{-1}$ when working at the same rate of $75\,\text{W}$, given that her total mass is $110\,\text{kg}$.

18 A racing car of mass $1500\,\text{kg}$, travelling at speed $v\,\text{m}\,\text{s}^{-1}$, experiences a resistance $R = 36v\,\text{N}$. The car travels down a slope at $2°$ to the horizontal. If the maximum power of its engine is $120\,\text{kW}$, calculate the maximum speed down the slope.

19 The object in question **12** is now dragged along a floor that is tilted upwards at $5°$ to the horizontal. All other given data stays the same. Calculate the speed up the slope and the engine power after moving 4 metres.

20 A train of mass $240\,000\,\text{kg}$ is pulled up a slope of $\arcsin\dfrac{1}{250}$ from rest. The resistance to motion is $2000g\,\text{N}$. The tractive force of its engine T varies with the distance s travelled, as in this table.

s (metres)	0	25	50	75	100	125	150
$T\,(\times 10^4\,\text{N})$	10	13	12	11	9	7	6

    a Use a numerical method to calculate an approximate value for the work done over 150 m.

    b Calculate the speed of the train after it has travelled 100 m, and calculate the power of its engine at this point, to the nearest watt.

    c Name two limitations to the model used in this problem.

    d What other numerical methods could you have used in part **a**? Which method would likely be more accurate? Explain your reasoning.

Fluency and skills

In many situations, it is not sensible to model **strings** as inextensible. You are now going to adopt a new model that assumes strings are **light** and **elastic** and that they recover their **natural length** after being stretched.

A **spring** is also modelled as light and elastic. As well as recovering its natural length after being stretched, it also recovers it after being compressed. A string cannot be compressed; it simply goes slack.

In reality, if the stretch goes beyond the **elastic limit**, the string or spring is permanently deformed and Hooke's law does not apply.

For a string or stretched spring, the difference between the stretched length l' and the natural length l is the **extension** x. For a compressed spring, the difference in length is called a **compression**, x

Hooke's law states that the **tension** T in an elastic string or spring is proportional to its extension x

$T \propto x$ where $x = l' - l$

So, $T = k \times x$ where the constant k is the **stiffness** of the string or spring.

Stiffness depends on the material that the string or spring is made from and also on its natural length. Stiffness k is equal to $\dfrac{\lambda}{l}$ where the constant λ is called the **modulus of elasticity**.

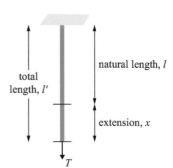

The units of λ are newtons.

Key point

Hooke's law gives $T = \dfrac{\lambda}{l} x$ where $x = l' - l$

To calculate the work required to create an extension X from its natural length, imagine a small extension δx when the tension is T.

The small amount of work required is force × distance $= T \times \delta x$. To calculate the total work needed, integrate for the whole extension. There are two methods.

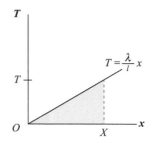

Using calculus

Work required $= \displaystyle\int_0^X T \, dx$

$= \dfrac{\lambda}{l} \displaystyle\int_0^X x \, dx = \dfrac{\lambda}{l} \left[\dfrac{x^2}{2} \right]_0^X$

$= \dfrac{1}{2} \times \dfrac{\lambda}{l} \times X^2 = \dfrac{\lambda}{2l} X^2$

Using a graph

Work required $= \displaystyle\int_0^X T \, dx$

$= $ Area of shaded triangle

$= \dfrac{1}{2} \times X \times T = \dfrac{1}{2} \times X \times \dfrac{\lambda}{l} \times X$

$= \dfrac{\lambda}{2l} X^2$

The work done to stretch a string or spring is stored as **elastic potential energy** (EPE) in the string or spring.

Key point

The elastic potential energy, EPE, stored in a string or spring when its natural length l is extended by a distance x is $\dfrac{\lambda}{2l}x^2$

This formula has the same shape as $\dfrac{1}{2}mv^2$ for kinetic energy.

To increase the extension from x_1 to x_2, the energy required is

$$\frac{\lambda}{2l}x_2^2 - \frac{\lambda}{2l}x_1^2 = \frac{\lambda}{2l}(x_2^2 - x_1^2) = \frac{\lambda}{2l}(x_2 + x_1)(x_2 - x_1)$$

Example 1

Two strings OA and OB are tied to an object O which is held on a smooth table between two fixed points A and B 1.0 m apart.

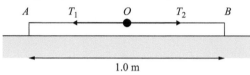

1.0 m

OA has natural length 0.4 m and modulus of elasticity 12 N.
OB has natural length 0.3 m and modulus of elasticity 18 N.

Find the extensions of the strings and their total EPE.

For horizontal equilibrium,　　　tension T_1 = tension T_2

> The object is held in equilibrium, so the two horizontal forces balance.

Hooke's law gives:

for OA,　$T_1 = \dfrac{\lambda_1}{l_1}x_1 = \dfrac{12}{0.4}x_1 = 30x_1$

> Use Hooke's law on OA and OB

for OB,　$T_2 = \dfrac{\lambda_2}{l_2}x_2 = \dfrac{18}{0.3}x_2 = 60x_2$

So　　$30x_1 = 60x_2$　　or　　$x_1 = 2x_2$　　(1)

> Use $T_1 = T_2$

Distance $AB = 1.0 = 0.4 + 0.3 + x_1 + x_2$

　　　　$0.3 = 2x_2 + x_2$　　　　(2)

The extensions are $x_2 = 0.1$ m and $x_1 = 0.2$ m

> Solve the simultaneous equations (1) and (2) to calculate x_1 and x_2

Total EPE $= \dfrac{\lambda_1}{2l_1}x_1^2 + \dfrac{\lambda_2}{2l_2}x_2^2$

　　　　$= \dfrac{12}{2 \times 0.4} \times 0.2^2 + \dfrac{18}{2 \times 0.3}0.1^2$

　　　　$= 0.6 + 0.3 = 0.9$ J

1 Five different strings A to E of natural length l and modulus of elasticity λ are extended by x.

Copy and complete this table, where T is the tension in each string.

	l, m	x, m	λ, N	T, N	EPE, J
A	2	0.5	8		
B	3	1	6		
C	1.5		15	4	
D	1.2	0.6		4.5	
E	2.5		10		1.28

2 A force of 15 N extends an elastic string by 0.2 metres. What will be the extension when the force is 30 N?

3 A force of 20 N compresses a spring by 5 cm. How far will the spring be compressed under a force of 10 N?

4 An elastic string of natural length 2 m with a modulus of elasticity of 50 N is extended by a force of 20 N.

Calculate

a   Its extension,   b   The work done by the force,

c   The elastic potential energy stored in the extended string.

5 A spring of natural length 1.2 m with $\lambda = 25$ N is compressed by a force of 15 N.

Calculate

a   How much it is compressed,

b   Its elastic potential energy when compressed.

6 An elastic string of natural length 1.5 m hangs from a ceiling with an object of mass 8 kg at its lower end. If $\lambda = 120$ N and the object hangs in equilibrium, calculate

a   The tension in the spring,   b   Its extension.

7 Two fixed points P and Q are 3 m apart on a smooth table. Two horizontal strings PO and QO are attached to an object O which sits between P and Q. The natural length and value of λ for PO are 0.9 m and 24 N and for QO are 0.6 m and 32 N. Calculate the extensions of the two strings and the tensions in them when O is in equilibrium.

8 Two springs OX and OY are tied to an object O which is in equilibrium on a smooth table between two fixed points X and Y, which are 1 m apart. OX has natural length 0.8 m and $\lambda = 15$ N. OY has natural length 1.2 m and $\lambda = 18$ N. Calculate the compressed lengths of the two springs and their total EPE.

9 Two light vertical springs AB and BC are joined at B with end A fixed to a horizontal table, and with C directly above B. A mass of 5 kg is attached to C, compresses both springs, and rests in equilibrium. The natural length and modulus of elasticity of AB are 0.4 m and 90 N, and those of BC are 0.2 m and 80 N. Find the distance between A and B and the total EPE stored in the springs.

10 Two vertical strings hang in parallel from a ceiling so that together they hold a mass m in equilibrium at their two lower ends. They have the same natural length l and their moduli of elasticity are λ_1 and λ_2 respectively. Show that the extension of each string is $\dfrac{mgl}{\lambda_1 + \lambda_2}$ and calculate the tension in each string in terms of m, g, λ_1 and λ_2

Strategy

To solve problems involving extensions in strings and springs

(1) Sketch a clear diagram to show all information.

(2) Use Hooke's law to calculate the energy stored in the string or spring.

(3) Write an energy equation that balances the changes in energy and any work done.

The principle of conservation of mechanical energy says that the total mechanical energy of a system remains constant, provided there is no external work done by resisting forces such as friction (or energy loss due to collisions).

This implies that any increase in one form of energy is balanced by a decrease in another form. When you write an energy equation, you may find it easier to think in terms of increases and decreases of KE, GPE and EPE rather than the total energy.

Example 2

A vertical elastic string, of natural length $AB = 1\,\text{m}$ with $\lambda = 49\,\text{N}$, has A fixed to a ceiling and B fixed to a mass of 2 kg. Take $g = 9.8\,\text{m}\,\text{s}^{-2}$.

a The mass is lowered gently until in equilibrium at point O. Calculate the extension when at O

It is then pulled down a further 20 cm to point P and released.

b Calculate its velocity v when passing O and the greatest height it reaches. Describe its subsequent motion.

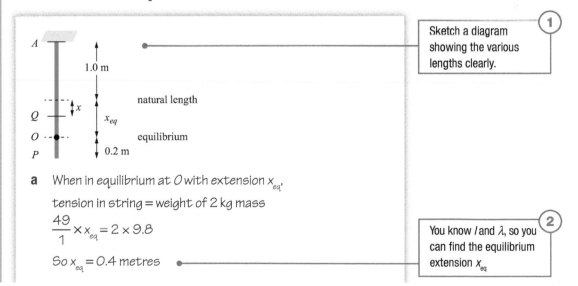

Sketch a diagram showing the various lengths clearly. **(1)**

a When in equilibrium at O with extension x_{eq},

tension in string = weight of 2 kg mass

$$\frac{49}{1} \times x_{eq} = 2 \times 9.8$$

So $x_{eq} = 0.4$ metres

You know l and λ, so you can find the equilibrium extension x_{eq} **(2)**

(Continued on the next page)

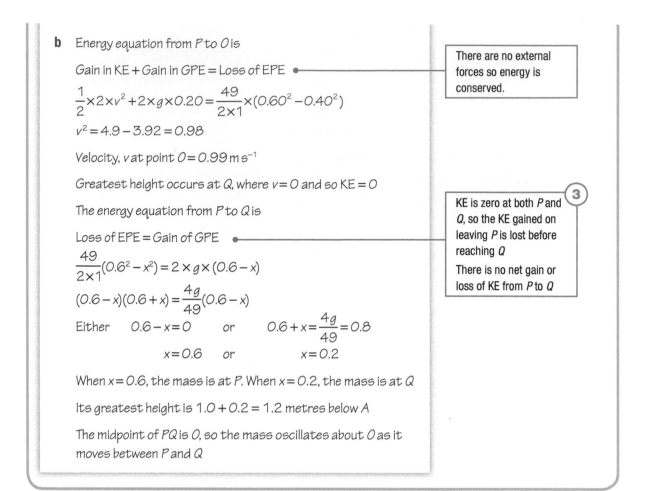

b Energy equation from P to O is

Gain in KE + Gain in GPE = Loss of EPE

$$\frac{1}{2} \times 2 \times v^2 + 2 \times g \times 0.20 = \frac{49}{2 \times 1} \times (0.60^2 - 0.40^2)$$

$$v^2 = 4.9 - 3.92 = 0.98$$

Velocity, v at point $O = 0.99\,\text{m s}^{-1}$

Greatest height occurs at Q, where $v = 0$ and so KE = 0

The energy equation from P to Q is

Loss of EPE = Gain of GPE

$$\frac{49}{2 \times 1}(0.6^2 - x^2) = 2 \times g \times (0.6 - x)$$

$$(0.6 - x)(0.6 + x) = \frac{4g}{49}(0.6 - x)$$

Either $\quad 0.6 - x = 0 \quad$ or $\quad 0.6 + x = \dfrac{4g}{49} = 0.8$

$\qquad\qquad x = 0.6 \quad$ or $\qquad\qquad x = 0.2$

When $x = 0.6$, the mass is at P. When $x = 0.2$, the mass is at Q

Its greatest height is $1.0 + 0.2 = 1.2$ metres below A

The midpoint of PQ is O, so the mass oscillates about O as it moves between P and Q

> There are no external forces so energy is conserved.

> ③ KE is zero at both P and Q, so the KE gained on leaving P is lost before reaching Q
>
> There is no net gain or loss of KE from P to Q

Exercise 1.2B Reasoning and problem-solving   `Full A Level`

In all questions, take the value of g to be $9.8\,\text{m s}^{-2}$ unless told otherwise.

1 How much work is done to stretch a spring from a length of 2 m to a length of 3 m, if its natural length is 1.5 m and its modulus of elasticity is 25 N?

2 A spring of natural length 0.5 m with $\lambda = 80$ N hangs from a ceiling. One end of an elastic string of natural length 0.6 m with $\lambda = 120$ N is tied to the spring's lower end. An object of mass 4 kg hangs in equilibrium from the lower end of the string.

Calculate

    a The tension in the spring and the string,   **b** The total extension,   **c** The total EPE.

3 Two thin light vertical springs, A and B, stand side by side with their lower ends fixed to a horizontal table. A mass of 5 kg is placed on their upper ends, compressing both springs so that the mass is in equilibrium. If the natural length and modulus of elasticity of A are 0.8 m and 80 N, and those of B are 1.5 m and 60 N, find

    a The compressed length of the springs,

    b The tension in each spring,

    c The total EPE in the springs.

4 a A model railway has a horizontal track which ends at a buffer made from a horizontal spring. In testing the spring when it is vertical, a mass of 2 kg compresses it by 2 cm. Subsequently, a truck of mass 0.5 kg runs into the buffer with a speed of 0.60 m s^{-1} Calculate the stiffness of the spring, $\dfrac{\lambda}{l}$. How much does the buffer compress in bringing the truck to rest?

b A vertical spring is fixed to a floor. It is compressed by a distance a when a mass m rests on it. If the mass falls from rest onto the spring from a height of $\dfrac{3a}{2}$, calculate the greatest compression of the spring in the subsequent motion.

5 A mass P of 2 kg on a rough horizontal table is tied to a horizontal elastic string, of natural length 2 m and $\lambda = 50$ N, whose other end is fixed to a point Q on the table. The mass is held such that $PQ = 3$ m. What is the initial acceleration of the mass and what is the distance PQ when the mass comes to rest, if the coefficient of friction $\mu = 0.6$?

6 A string of natural length l and modulus λ has an initial extension of x_1. It is stretched further until its extension is x_2. Prove that the work done to increase the extension from x_1 to x_2 equals the product of the increase in the extension and the average (mean) of the initial and final tensions.

7 A horizontal spring, of natural length 2 m and modulus of elasticity 50 N, has one end fixed to point A on a horizontal table. The other end is tied to a mass of 0.2 kg. The spring is then compressed 0.4 metres and released. How far does the mass travel before coming to instantaneous rest, if the table is

a Smooth, **b** Rough with $\mu = 0.4$?

8 A mass of 1 kg is tied to the lower end of a vertical elastic string, of natural length $AB = 1.5$ m with $\lambda = 49$ N. The other end, A, is fixed to a ceiling.

a When the mass is in equilibrium at point P, what is the extension in the string?

b The mass is now pulled down to point Q 30 cm below P and released.
Find its velocity v when it reaches P and the greatest height to which it rises above P. Describe its subsequent motion.

c The mass is pulled down further to point R so that it is 50 cm below P. When released, how close does it get to point A?

d If the string was replaced by a spring with the same natural length AB and modulus λ, how would your answer to part **b** be different? How close does it now get to A?

9 A light elastic rope of natural length 25 m with $\lambda = 1200$ N has its ends tied to a bungee jumper of mass 60 kg and to a bridge. The jumper falls from rest off the bridge and reaches a speed of v m s^{-1} after dropping a distance x m.

a Show that $30v^2 = 1788x - 24x^2 - 15\,000$ provided $x > k$. State the value of k and calculate the maximum value of v

b Calculate the maximum value of x

c Suggest two ways in which the model in this problem could be refined to make it more realistic.

10 An object of 4 kg lies on a plane which slopes at 30° to the horizontal. A spring, of natural length 1.5 m with $\lambda = 40$ N, is attached to the top of the plane with the other end attached to the object.

   a   If the plane is smooth, calculate the length of the spring when the object rests in equilibrium.

   b   If the plane is rough with $\mu = 0.2$, calculate the length of the section of the plane where the object can rest in equilibrium.

   c   Explain how a smooth plane would affect your answer for part **b**.

11 An elastic string of natural length 3 m hangs vertically and is stretched to a length of 4.5 m by a mass of 15 kg. The same mass is now attached to the midpoint of the same string which is stretched horizontally between two points X and Y which are 4 metres apart.

   a   On release, the mass falls vertically from rest. Calculate its velocity after it falls 1.5 m.

   b   Verify that, if the mass is gently lowered until it hangs in equilibrium, its distance below the horizontal through X and Y is approximately 1.1 m.

12 A light rod AB of length 2 m is hinged to a vertical wall at A with B higher than A. A light elastic string, of natural length 1 m with $\lambda = 100$ N, is stretched from B to point C on the wall 2 m above A

A mass of m kg is fixed to B

   a   If angle $CAB = 60°$, calculate the value of m when the mass is in equilibrium.

   b   If B is placed to coincide with C and the mass m is released from rest in this position so that AB rotates about A, what is the maximum value of angle CAB in the subsequent motion?

   c   Describe how this model could be refined to make the answers more realistic.

Chapter summary

- Work done by a constant force F moving a distance s in its direction is $F \times s$ and, for a variable force, work done $= \int F \, ds$
- Energy is gained by an object when work is done on it. Mechanical energy is either kinetic energy or potential energy. The units of energy and work are joules.
- Kinetic energy, $\text{KE} = \dfrac{1}{2} mv^2$, relates to a body's motion.
- Gravitational potential energy, $\text{GPE} = mgh$, relates to a body's vertical height in Earth's gravity.
- Power is the rate of doing work. For a constant force F moving with a speed v, power $= F \times v$. The unit of power is watts.
- When there is no external input (or loss) of work or energy, the total energy of a system is conserved. So, gains in one form of energy are balanced by losses in another form of energy.

  Full A Level

- Elastic potential energy, $\text{EPE} = \dfrac{\lambda}{2l} x^2$ relates to an extension of an elastic string or spring, or a compression of a spring.
- Hooke's law states that the tension T in a string or spring is proportional to its extension (or, for a spring, its compression), x, where $T = \dfrac{\lambda}{l} x$

Check and review

You should now be able to...	Try Questions
✔ Analyse problems using formulae for work, energy and power.	1–13
✔ Use Hooke's law to solve problems involving strings and springs.	10, 11, 14

In all questions, take the value of g to be $9.8 \, \text{m s}^{-2}$ unless told otherwise.

1   A car has a mass of 1200 kg. It starts from rest with a constant tractive force and reaches a speed of $24 \, \text{m s}^{-1}$ after travelling 320 metres. Assuming that resistances to motion are negligible, use an energy equation to calculate the tractive force.

2   Reindeer pull a sleigh of mass 300 kg horizontally over 100 m on ice from rest with a constant acceleration of $0.5 \, \text{m s}^{-2}$. Assuming negligible resistance to motion, calculate

   a   The horizontal force pulling the sleigh,   **b**   The work done by the reindeer,

   c   The gain in KE of the sleigh,   **d**   The power of the reindeer after 100 m.

3 A box of mass 200 kg is winched, with constant force, 12 metres vertically for 10 seconds from rest by a rope with constant tension. Calculate

a Its acceleration,   **b** Its final velocity,

c The gain in its KE,   **d** The gain in its GPE,

e The work done by the winch,   **f** The tension in the rope,

g The power of the winch after 10 seconds.

4 A car of mass 1000 kg is travelling at $10\,\text{m s}^{-1}$ when the driver decides to stop. She brakes and keeps braking for the entire distance of 80 m to come to rest. She also disengages the engine after the car has travelled the first 20 metres. If the car's engine has constant a tractive force of 120 N, calculate the resisting force of the brakes, assuming that it is constant.

5 A fountain pumps 1500 litres of water 2 metres vertically every minute and expels it at a speed of $5\,\text{m s}^{-1}$. Assuming no loss of energy, find the power of the pump. The density of water is 1 kg per litre.

6 A cyclist of total mass 90 kg has a speed of $10\,\text{m s}^{-1}$ at point P at the top of a hill. She free-wheels downhill to point Q, a vertical distance of 50 m below P, reaching Q with a speed of $24\,\text{m s}^{-1}$. The distance PQ by road is 360 m. There is a constant resistance to motion; calculate its value.

7 A 2 kg mass is dragged along a rough horizontal floor from rest for a distance s metres by a variable force F where $F = 24 + 2s - s^2$ newtons. If the coefficient of friction μ is $\dfrac{2}{7}$, what is the speed of the mass after moving 3 metres? What is the power of the force at this point?

8 The resistance to a car's motion when moving with speed v is kv, where k is a constant, and the maximum power of its engine is P. Its maximum speed down a slope is V. Its maximum speed up the same slope is $\dfrac{1}{2}V$. Prove that $V = \sqrt{\dfrac{2P}{k}}$

9 Force F can vary with distance s. Find the work needed to lift a satellite (of mass 1000 kg) 4000 km above the Earth's surface when the gravitational force FN on the satellite is given by $F = \dfrac{k}{s^2}$ where s m is its distance from the centre of the Earth. Take $k = 4 \times 10^{17}$ and the Earth's radius as 6400 km.

10 A mass of 12 kg hangs from an elastic string of natural length 1.5 m whose upper end is fixed to a ceiling. If $\lambda = 120\,\text{N}$ and the mass is in equilibrium, calculate

a The tension in the string,

b Its extension,

c The EPE stored in it.

11 Two fixed points X and Y are 2 m apart on a smooth table. Two horizontal strings OX and OY are attached to an object O which lies between X and Y. The natural length and modulus of elasticity for OX are 0.6 m and 16 N and for OY are 0.4 m and 20 N. Find the extensions of the two strings and the tensions in them when O is in equilibrium.

12 A lorry of mass 3 tonnes travels up a slope at an angle of $\arcsin \dfrac{1}{140}$ to the horizontal and its speed increases steadily, under a constant tractive force, from $15\,\text{m s}^{-1}$ to $25\,\text{m s}^{-1}$ over a distance of 500 metres. Assuming there is no resistance to motion, find the total increase in its energy, the tractive force of its engine and its final power output.

13 The resistance to a car of mass $800\,\text{kg}$ moving at $v\,\text{m s}^{-1}$ is given by $34 + 5v^2\,\text{N}$. When travelling up a slope at $\arcsin \dfrac{1}{140}$ to the horizontal, its engine is working at its maximum power of $40\,\text{kW}$. Show that the maximum speed $V\,\text{m s}^{-1}$ up the slope satisfies a cubic equation with a solution of $19.7\,\text{m s}^{-1}$. When free-wheeling down the same slope, what is its maximum speed?

14 An elastic string has a natural length $4\,\text{m}$ and a modulus of elasticity of $20g\,\text{N}$. The string is stretched horizontally between two points A and B, which are 6 metres apart, and a mass of $15\,\text{kg}$ is attached to its midpoint.

   a The mass drops vertically after being released from rest. Calculate its velocity after it falls $3\,\text{m}$.

   b If, instead, the mass is gently lowered until it hangs in equilibrium, verify that its distance below the midpoint of AB is approximately $1.76\,\text{m}$.

History

Robert Hooke was a remarkable scientist who lived in the 17th century. He lends his name to Hooke's law of elasticity, devised in 1660, but he carried out research in a wide range of fields. Perhaps most notably, he illustrated his observations through a microscope in the book *Micrographia: or Some Physiological Descriptions of Minute Bodies Made by Magnifying Glasses*. His amazing drawings were the first major publication of the Royal Society in 1665. Far from the world of microscopes, he also explored the stars and planets with telescopes, drawing some very detailed sketches of Mars.

Investigation

A section of a rollercoaster can be modelled as a ball, with mass 100 kg and negligible radius, rolling around the inside of a vertical circle with radius 10 m. The ball needs to be travelling at 10 m s⁻¹ at the top of the circle for it to make a complete loop-the-loop without leaving the track. Investigate what speed the ball has to be travelling as it enters the circle. How would the situation change if the ball were fixed to the circle, so that it couldn't fall off?

Note

The SI unit for energy is the same as the unit of work – the joule (J). This is named in honour of James Prescott Joule. 1 joule is the work done when lifting an object with a weight of 1 newton by 1 metre.

Research

Look up different landing vehicles that have visited other planets or moons in the Solar System. How much do they weigh, how much energy do they need to leave the surface, and in what way is the energy affected by the size of the moon or planet?

Investigation

In a bungee jump, people are tied to an elastic rope of a carefully calculated length, so that when they fall their head just dips into the water below. Investigate how the length of the rope is varied, based on the weight of the jumper as a fraction of the typical male adult.

In all questions, take the value of g to be $9.8\ \mathrm{m\,s^{-2}}$ unless told otherwise.

1 A man raises a load of 20 kg through a height of 6 m using a rope and pulley. His maximum power is 180 W. Assuming the rope and pulley are light and smooth,

    a How much work does he do in completing the task, **[2 marks]**

    b What is the shortest time in which he could complete it? **[2]**

    c **i** Explain how your results might be affected if the rope had significant mass. **[2]**

        ii Is it reasonable to suppose that the maximum power of 180 W can be applied throughout the task? Explain your answer. **[2]**

2 A child of mass 25 kg moves from rest down a slide. The total vertical drop in height is 4 m.

    a Assuming that friction is negligible, calculate the speed of the child at the bottom of the slide. **[3]**

    b In fact, the child reaches the bottom travelling at $6\ \mathrm{m\,s^{-1}}$. The length of the slide is 6 m. Calculate

        i The work done against friction, **[2]**

        ii The frictional force (assuming it to be constant). **[2]**

3 A horizontal force is applied to a 6 kg body so that it accelerates uniformly from rest and moves across a horizontal plane against a constant frictional resistance of 30 N. After it has travelled 16 m, it has a speed of $4\ \mathrm{m\,s^{-1}}$. Calculate

    a The applied force, **[2]**

    b The total work done by the force. **[2]**

4 Two elastic strings, *AB* and *BC*, are joined at *B*, and the other ends are fixed to points *A* and *C* on a smooth horizontal table. *AB* has natural length 85 cm, and modulus of elasticity 45 N. *BC* has natural length 45 cm and modulus of elasticity 65 N. The distance *AC* is 2.4 m. Calculate the stretched lengths of the strings. **[4]**

Full A Level

5   Particles A and B, of mass 0.5 kg and 1.5 kg respectively, are connected by a light inextensible string of length 2.5 m. Particle A rests on a smooth horizontal table at a distance of 1.5 m from its edge. The string passes directly over the smooth edge of the table and particle B hangs suspended. The system is held at rest with the string taut and then released. Calculate the speed of A when it reaches the edge of the table. [3]

6   A pump, working at 3 kW, raises water from a tank at 1.2 m^3 min^{-1} and emits it through a nozzle at 15 m s^{-1}. Calculate the height through which the water is raised. (Density of water = 1000 kg m^{-3}) [4]

7   A car of mass 800 kg has a maximum speed of 75 km h^{-1} up a slope against a constant resistance of 500 N. The slope is inclined at $\sin^{-1}\dfrac{1}{40}$ to the horizontal. Calculate the power of the engine. [4]

8   The resistance to motion of a car is proportional to its speed. A car of mass 1000 kg has a maximum speed of 45 m s^{-1} on the level when its power is 8 kW. Calculate its acceleration when it is travelling on the level at 20 m s^{-1} and its engine is working at 6 kW. [6]

9   A ball of mass 500 g is fastened to one end of a light, elastic rope, whose natural length is 3 m and whose modulus of elasticity is 90 N. The other end of the rope is fastened to a bridge. The ball is held level with the fixed end, and is released from rest.

   a   Calculate the speed of the ball when the rope just becomes taut. [2]

   b   How far below the bridge is the lowest point reached by the ball? [4]

   c   i   What is the main additional modelling assumption you have made in your solution? [1]

       ii In what way would the result be affected if that assumption were false? [1]

10 A particle of mass 2 kg is suspended from a point A on the end of a spring of natural length 1 m and modulus of elasticity 196 N.

   a   Calculate the length of the spring when the particle hangs in equilibrium. [3]

   The particle is now pulled down a distance of 0.5 m and released from rest.

   b   Calculate the distance below A at which the particle next comes instantaneously to rest, assuming that the spring can compress to that point without the coils touching. [5]

   c   Calculate the highest position reached by the particle if, instead of the spring, you had used an elastic string of the same natural length and modulus of elasticity. [2]

11 A particle of mass 2 kg is attached to one end of an elastic string of natural length 2 m and modulus of elasticity 40 N. The other end of the string is attached to a fixed point O on a rough horizontal plane. The coefficient of friction between the particle and the plane is 0.5. The particle is projected from O along the plane with initial velocity v m s^{-1}. The particle returns and comes to rest exactly at O. Calculate

   a   Its furthest distance from O,   [7]

   b   The value of v [2]

12 A catapult is made by fastening an elastic string of natural length 10 cm to points A and B, a distance of 6 cm apart. The modulus of elasticity of the string is 5 N. A stone of mass 10 g is placed at the centre of the string, which is then pulled back until the stone is 25 cm from the centre of AB.

   a   Calculate the greatest speed reached by the stone when it is released.   [6]

   b   Would the result be affected by the direction in which the stone is fired? Explain your answer.   [2]

13 A particle of mass 2 kg is attached to one end of an elastic string of natural length 1.2 m and modulus of elasticity 240 N. The other end of the string is fixed to a point A on a rough, horizontal plane. The particle is held at rest on the plane with the string stretched and is then released. The particle just reaches A before coming to rest. The coefficient of friction between the particle and the plane is 0.5

   a   Calculate the initial extension of the string.   [6]

   b   Calculate the speed of the particle at the moment when the string went slack.   [4]

14 A car working at a rate P watts has a maximum speed V m s^{-1} when travelling on the level against a resistance proportional to the square of its speed. At what rate would the car have to work to double its maximum speed in terms of P?   [4]

15 A car of mass 1 tonne is towing a trailer of mass 400 kg on a level road. The resistance to motion of the car is 400 N and of the trailer is 300 N. At a certain instant, they are travelling at 10 m s^{-1} and the power output of the engine is 10.5 kW. Calculate the tension in the coupling between the car and the trailer.   [4]

16 The frictional resistances acting on a train are $\dfrac{1}{100}$ of its weight. Its maximum speed up an incline of $\sin^{-1}\dfrac{1}{80}$ is 48 km h^{-1}. Calculate its maximum speed on the level, in km h^{-1}.   [7]

17 A cyclist and her cycle have a combined mass of 80 kg. The resistance to motion is proportional to the speed. On the level, she can travel at a maximum speed of $10\,\text{m}\,\text{s}^{-1}$, and she can free-wheel down an incline of angle θ at a steady speed of $14\,\text{m}\,\text{s}^{-1}$. Calculate the maximum speed at which she can go up the same incline. **[6]**

18 A block of mass 2 kg rests on a rough plane which is inclined at 30° to the horizontal. The block is attached to a point at the top of the plane by means of an elastic string of natural length 2 m and modulus of elasticity 100 N. The coefficient of friction between the block and the plane is 0.25. Calculate the distance between the lowest and highest positions in which the block will rest in equilibrium. **[6]**

19 A lorry of mass 5.5 tonnes is travelling up a hill inclined at $\sin^{-1}\dfrac{1}{50}$ to the horizontal. When its speed is $v\,\text{m}\,\text{s}^{-1}$, the resistance to motion is $kv\,\text{N}$, where k is a constant. The lorry's power is 12 kW and it is travelling at its maximum speed of $10\,\text{m}\,\text{s}^{-1}$.

a Calculate the value of k **[3]**

The lorry reaches the top of the hill and moves onto a level road.

b Show that its initial acceleration is $0.196\,\text{m}\,\text{s}^{-2}$ **[2]**

c Calculate its new maximum speed. **[2]**

20 A car of mass 900 kg moves against a resistance which is proportional to its speed. Its maximum power output is 6 kW and on a level road its maximum speed is $40\,\text{m}\,\text{s}^{-1}$. Calculate its maximum speed up an incline whose angle to the horizontal is θ, where $\sin\theta = \dfrac{1}{30}$ **[6]**

21 A body of mass 10 kg is at rest on a rough horizontal surface. The coefficient of friction between the body and the surface is 0.5. The body is then pulled for a distance of 20 m, in a straight line, by a light inelastic string inclined at 30° to the horizontal. The tension in the string is 50 N. Calculate

a The work done by the tension, **[2]**

b The final speed of the body. **[3]**

c By increasing the tension in the string, the final speed of the body can be increased. Give two reasons why it cannot be increased indefinitely. **[2]**

Momentum and collisions

2

Forensic scientists use the principles of conservation of momentum, and momentum change during impacts, to construct models of collisions between two vehicles. Data from the accident scene, such as measurements of tyre skid marks and vehicle compression, are used to perform calculations that can reconstruct collisions and make an expert judgement of what really happened. These calculations are used in court cases to determine who was really at fault during a collision, and so it's vital that the calculations involved are accurate and a true reflection of reality.

Vehicle collisions are just one area in which principles relating to momentum are vitally important. Other areas include space engineering and rocket science, the design of theme park rides, and railway engineering.

Orientation

What you need to know	What you will learn	What this leads to
Maths Ch7 • Equations of motion	• To use the momentum equation and Newton's equation appropriately.	**Careers** • Forensic science. • Automotive testing. • Mechanical engineering.
Maths Ch8 • Resolving forces • Newton's laws of motion	• To find the size of an impulse for constant and variable forces.	
Mechanics Ch1 • Kinetic energy	• To apply these concepts to problems in 2 dimensions. A Level	

p.3

Fluency and skills

When a particle of mass m moves with velocity v, its **momentum** is equal to the mass multiplied by the velocity. If it is acted on by a constant force F for a time t, then the **impulse** acting on the particle is equal to the force multiplied by the time during which the force acts.

Key point

$$\text{Momentum} = m \times v \qquad \text{Impulse} = F \times t$$

> The units of momentum and impulse are the same. They are newton seconds (N s).

F and v are vectors, so momentum and impulse are also vectors.

If the force F is constant and the particle's velocity increases from u to v, then its acceleration is $a = \dfrac{v-u}{t}$

Newton's 2nd law gives $F = ma = \dfrac{m(v-u)}{t} = \dfrac{mv-mu}{t}$

Key point

Impulse of force F = Change in momentum
$$Ft = mv - mu$$

When bodies A and B with masses m_1 and m_2 and initial velocities u_1 and u_2 collide, forces act on the bodies for a short time t

> Note that velocity is a vector, so u and v can be negative and therefore in the opposite direction to the arrow.

> In most situations, colliding objects will separate and move with different velocities v_1 and v_2 after the collision. If, however, they stick together and begin to move with the same velocity, they are said to **coalesce**.

If the force acting on B is F, then Newton's 3rd law states that an equal and opposite force $-F$ acts on A

After the collision, A and B move with velocities v_1 and v_2

For A,　　$-Ft = m_1v_1 - m_1u_1$

For B,　　$Ft = m_2v_2 - m_2u_2$

So　　$-(m_1v_1 - m_1u_1) = m_2v_2 - m_2u_2$

Rearrange

$$m_1u_1 + m_2u_2 = m_1v_1 + m_2v_2$$

Total initial momentum = Total final momentum

Key point

The **principle of conservation of (linear) momentum** states that, when no *external* forces are present, the total momentum of a system of particles is unchanged by collisions between them.

Example 1

A body P of mass 1 kg moving with velocity $5\,\text{m s}^{-1}$ collides directly with another body Q of mass 2 kg moving towards P with velocity $4\,\text{m s}^{-1}$. After the collision, Q is at rest. Find the final velocity of P

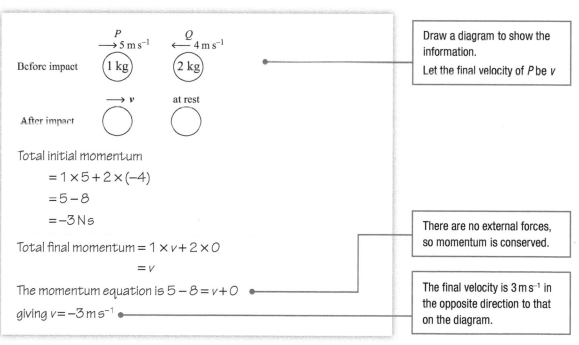

Before impact

P $\longrightarrow 5\,\text{m s}^{-1}$　Q $\longleftarrow 4\,\text{m s}^{-1}$

1 kg　　2 kg

After impact

$\longrightarrow v$　at rest

Draw a diagram to show the information.

Let the final velocity of P be v

Total initial momentum

$= 1 \times 5 + 2 \times (-4)$

$= 5 - 8$

$= -3\,\text{Ns}$

Total final momentum $= 1 \times v + 2 \times 0$

$= v$

There are no external forces, so momentum is conserved.

The momentum equation is $5 - 8 = v + 0$

giving $v = -3\,\text{m s}^{-1}$

The final velocity is $3\,\text{m s}^{-1}$ in the opposite direction to that on the diagram.

Note that in Example 1, the final velocity of P is *negative*, so it's moving backwards (to the left) compared to its initial movement. This is, in fact, the only possible direction given that Q ended at rest, and the bodies cannot pass through each other.

It's useful in questions like this to check the direction is sensible after you've done your calculations – if objects have passed through each other, then something has likely gone wrong.

1

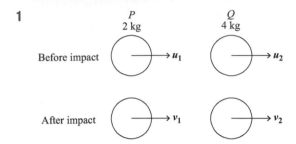

P
2 kg

Q
4 kg

Before impact $\rightarrow u_1$ $\rightarrow u_2$

After impact $\rightarrow v_1$ $\rightarrow v_2$

Two particles P and Q, with masses $2\,\text{kg}$ and $4\,\text{kg}$ respectively, move along a straight line. They collide directly, travelling with velocities u_1 and u_2 before impact and with velocities v_1 and v_2 after impact. Calculate the values, x, of the unknown velocities in this table. All velocities are given in m s^{-1}

	u_1	u_2	v_1	v_2
a	3	2	2	x
b	6	1	x	3.5
c	5	4	4	x
d	10	−2	x	4
e	1	−4	−4	x

2 Two particles Y and Z, with masses m_1 and $m_2\,\text{kg}$ respectively, move along a straight line with velocities u_1 and $u_2\,\text{m s}^{-1}$. They impact directly, coalesce, and then travel with velocity v. Calculate v in each case. All velocities are given in m s^{-1}

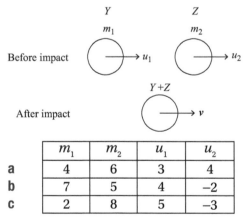

Y
m_1

Z
m_2

Before impact $\rightarrow u_1$ $\rightarrow u_2$

$Y+Z$

After impact $\rightarrow v$

	m_1	m_2	u_1	u_2
a	4	6	3	4
b	7	5	4	−2
c	2	8	5	−3

3 Particle A of mass $4\,\text{kg}$ moves at $5\,\text{m s}^{-1}$ towards particle B which is at rest. After a direct impact, A is at rest and B moves at $2\,\text{m s}^{-1}$. What is the mass of B?

4 A bullet of mass 40 grams travelling at $80\,\text{m s}^{-1}$ hits and coalesces with a stationary wooden target of mass $2\,\text{kg}$ which is free to move. Find their common speed v immediately after impact.

Reasoning and problem-solving

Strategy

To solve collision problems involving the conservation of momentum:

(1) Draw and label a diagram showing the situation before and after the collision.

(2) Check that there are no external forces involved, so that momentum is conserved.

(3) Write momentum equations, ensuring correct signs for velocities, and solve them.

Example 2

> The * in the diagram indicates a collision.

Three particles A, B and C, with masses 1 kg, 4 kg and 12 kg respectively, are positioned in a straight line. Particles B and C are at rest and particle A is moving towards B with a speed of 10 m s^{-1}. After A and B collide, particle A rebounds backwards and B moves towards C with twice the speed of A. After B and C collide, they move in opposite directions with the same speed.

Show that there are no more collisions between the particles.

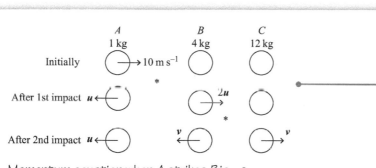

Draw a clearly-labelled diagram showing the velocities of the particles at different stages.

There are no external forces, so momentum is always conserved.

Momentum equation when A strikes B is

$$1 \times 10 + 0 = 1 \times (-u) + 4 \times 2u$$

$$\Rightarrow \quad 7u = 10$$

$$\Rightarrow \quad u = 1\frac{3}{7} \text{ m s}^{-1}$$

Momentum equation when B strikes C is

$$4 \times 2u + 0 = 4 \times (-v) + 12 \times v$$

$$\Rightarrow \quad 8v = 8u$$

$$\Rightarrow \quad v = u = 1\frac{3}{7} \text{ m s}^{-1}$$

After these two collisions, A and B are moving in the same direction with the same speed and in the opposite direction to C. Hence, there are no further collisions.

Write a momentum equation for each collision. Note the negative signs.

1   A hammer of mass 10 kg travelling vertically downwards with speed $24\,\mathrm{m\,s^{-1}}$ strikes the top of a vertical post of mass 2 kg and does not rebound. Find the speed of the hammer and post immediately after impact.

2   An arrow, of mass 250 grams and travelling at $25\,\mathrm{m\,s^{-1}}$, hits directly and coalesces with a stationary target of mass 9.0 kg which can move freely. Calculate their velocity immediately after impact.

3   Two cars of mass 900 kg and 1200 kg are travelling in opposite directions on a straight road with speeds of $135\,\mathrm{km\,h^{-1}}$ and $90\,\mathrm{km\,h^{-1}}$ respectively when they collide head-on and lock together. What is their speed in $\mathrm{km\,h^{-1}}$ immediately after impact?

4   A bullet of mass 40 grams is fired horizontally with a speed of $600\,\mathrm{m\,s^{-1}}$ into a target of mass 2 kg hanging from a vertical string. The bullet becomes embedded in the target. Find their common speed immediately after the impact.

5   Two railway trucks with masses 2 tonnes and 3 tonnes are travelling on the same line at speeds of $5\,\mathrm{m\,s^{-1}}$ and $2\,\mathrm{m\,s^{-1}}$ respectively. They collide and become coupled together. Find their common velocity after the collision if they were initially travelling

   a   In the same direction,

   b   In opposite directions.

6   When a gun is fired, the bullet shoots forwards and the gun is free to recoil backwards. The bullet has mass 50 grams and the gun has mass 2 kg. If the speed of the bullet is $250\,\mathrm{m\,s^{-1}}$, find the speed at which the gun recoils.

   Given that the explosion creates a force, explain why the principle of conservation of momentum can still be used.

7   A gun fires a shell of mass 10 kg. When travelling horizontally at $120\,\mathrm{m\,s^{-1}}$, it explodes into two pieces which also initially travel horizontally. Immediately after the explosion, one piece of mass 4 kg travels backwards with speed $40\,\mathrm{m\,s^{-1}}$. The other piece continues forward with a speed v. Calculate the value of v

8   Three spheres A, B and C, with equal radius and masses 5 kg, 4 kg and 3 kg respectively, lie in a straight line with B and C at rest on a smooth horizontal surface. Sphere A moves towards B with a speed of $18\,\mathrm{m\,s^{-1}}$. On impact, A and B coalesce and move towards C. On impact, they coalesce with C and move with velocity v

   a   Calculate v

   b   Name one limitation to the model adopted in this problem.

9 Three particles X, Y and Z, with masses 4 kg, 2 kg and 6 kg respectively, lie in a straight line, on a smooth horizontal surface, with X and Z at rest. Y moves towards Z at 10 m s^{-1} and collides with Z. After impact, Y and Z both move with speed u in opposite directions. Y now collides and coalesces with X and they continue with speed v

   a   Calculate u and v and explain why there is or is not another collision.

   b   List some of the assumptions that must be made to solve this problem.

10 Three particles R, S and T, of mass 6 kg, 3 kg and 5 kg respectively, lie in a straight line in the order $R\,S\,T$. R and S are joined by a light, inextensible, slack string. R and T are at rest. S moves away from R with a speed of 10 m s^{-1} and the string becomes taut before S reaches T

   a   Calculate the speed of R after the string becomes taut.

   b S now collides with T and comes to rest. Calculate the speed of T after this collision. Explain why there is a further collision between R and S

   c   Explain one way in which the model used in your solution could be made more realistic.

11 Three particles A, B and C, of mass 2 kg, 3 kg and 1 kg respectively, are in line with B and C at rest and A moving towards B with speed u. The collision between A and B brings A to rest. B now collides with C. After this collision, B and C move in the same direction, with C moving twice as fast as B. Find the final speeds of B and C in terms of u

12 Particle A of mass 10 kg has speed 5 m s^{-1}. It collides directly with particle B of mass m kg moving in the opposite direction at 2 m s^{-1}. After the collision, A continues in the same direction with a speed of 3 m s^{-1}. Show that, if there are no further collisions, $m \le 4$

Fluency and skills

When two bodies with known masses and velocities collide and rebound, their velocities after impact are unknown. You need two equations to find them. One is the momentum equation. The other is given by **Newton's law of restitution**. It involves the speed at which the gap between two bodies changes.

Key point

$$\frac{\text{Speed of separation}}{\text{Speed of approach}} = e$$

The constant e is the **coefficient of restitution**. It depends on the **elasticity** of the bodies and has a value in the range $0 \leq e \leq 1$

$e = 0$ if the impact is **inelastic** (that is, there is no rebound and the bodies coalesce).

$e = 1$ if the impact is **perfectly elastic**.

If you say the velocities of the objects before impact are u_1 and u_2, and the velocities after impact are v_1 and v_2, then you can use the following equation.

Key point

$$e = \frac{v_2 - v_1}{u_1 - u_2}$$

The **speed of approach** is the component of the speed along the line of impact at which the gap decreases before impact.

The **speed of separation** is the component of the speed along the line of impact at which the gap increases after impact.

A golf ball has $e \approx 0.8$ for an impact with a hard object, like a golf club.

Example 1

A particle P with mass 4 kg and velocity 20 m s^{-1} collides directly with particle Q with mass 12 kg travelling with velocity 4 m s^{-1} in the opposite direction.

Given $e = \frac{1}{2}$, calculate the velocities v_1 and v_2 after impact.

(Continued on the next page)

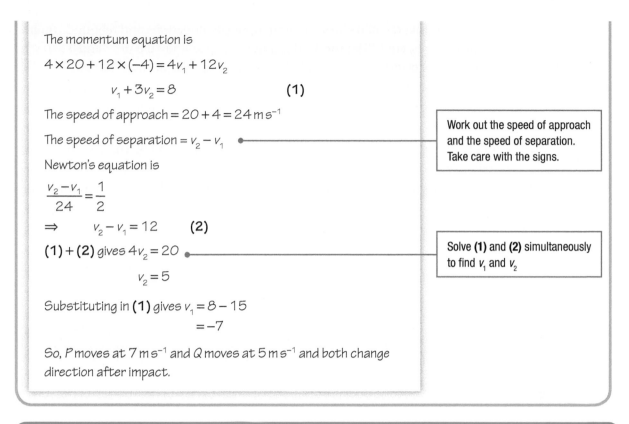

The momentum equation is

$$4 \times 20 + 12 \times (-4) = 4v_1 + 12v_2$$

$$v_1 + 3v_2 = 8 \qquad\qquad (1)$$

The speed of approach $= 20 + 4 = 24\,\mathrm{m\,s^{-1}}$

The speed of separation $= v_2 - v_1$

> Work out the speed of approach and the speed of separation. Take care with the signs.

Newton's equation is

$$\frac{v_2 - v_1}{24} = \frac{1}{2}$$

$$\Rightarrow \qquad v_2 - v_1 = 12 \qquad (2)$$

$(1) + (2)$ gives $4v_2 = 20$

> Solve (1) and (2) simultaneously to find v_1 and v_2

$$v_2 = 5$$

Substituting in (1) gives $v_1 = 8 - 15$

$$= -7$$

So, P moves at $7\,\mathrm{m\,s^{-1}}$ and Q moves at $5\,\mathrm{m\,s^{-1}}$ and both change direction after impact.

Exercise 2.2A Fluency and skills

1 Two particles P and Q, with masses $2\,\mathrm{kg}$ and $4\,\mathrm{kg}$ respectively, move in a straight line and collide directly. They travel with velocities u_1 and u_2 before impact and with velocities v_1 and v_2 after impact. In each case, calculate their velocities after impact.

	u_1	u_2	e
a	5	3	$\frac{1}{2}$
b	6	-2	$\frac{3}{4}$
c	1	10	$\frac{2}{3}$
d	5	-4	$\frac{1}{3}$
e	8	-2	$\frac{1}{4}$

2 Two particles R and S, with masses $3\,\mathrm{kg}$ and $6\,\mathrm{kg}$ respectively, move in a straight line and collide directly. They travel with velocities u_1 and u_2 before impact and with velocities v_1 and v_2 after impact. In each case, calculate the value of v_2 and the coefficient of restitution e

	u_1	u_2	v_1
a	16	4	2
b	5	2	2.5
c	-6	4	2
d	4	-1	-2

3 Two dodgem cars at a fairground collide head-on from opposite directions and rebound. Their masses, including passengers, are 125 kg and 150 kg, and their speeds are 0.6 m s^{-1} and 0.5 m s^{-1} respectively. The coefficient of restitution is 0.3. Calculate their speeds just after impact.

4 An ice-hockey puck of mass 800 grams skims across the ice at 15 m s^{-1}. It hits an identical puck, which is stationary, and both move on in the same straight line. If $e = 0.6$, find their velocities immediately after the collision.

5 A smooth 4 kg sphere drops from rest onto a horizontal floor from a height of 3 m. If $e = 0.6$, find its velocity immediately after impact and the loss of kinetic energy due to the impact.

6 A golf ball of mass 46 grams drops 2.5 m from rest onto a solid horizontal floor and rebounds to a height of 2.0 m. Find the coefficient of restitution, e, and the loss of kinetic energy on impact.

7 Two toy trains travel, with no driving force, on the same track in the same direction and collide. The front and rear trains have masses of 0.15 kg and 0.24 kg and initial speeds of 5.0 cm s^{-1} and 8.0 cm s^{-1} respectively. After impact, the rear train has a speed of 6.0 cm s^{-1} in the same direction.

    a Find the value of e

    b Find the loss of KE during the impact.

Reasoning and problem-solving

Strategy

To solve a collision problem involving elasticity:

(1) Draw and label a diagram showing the situation before and after the collision.

(2) Check whether there are any external forces and whether momentum is conserved.

(3) Write the momentum equation and Newton's equation, taking care with signs, and solve.

Example 2

Two spheres P of mass 1 kg and Q of mass 3 kg lie on a smooth horizontal plane with the line PQ at right-angles to a vertical wall.

P moves at 10 m s^{-1} and collides directly with Q at rest.

Q then hits the wall and rebounds.

a If the coefficient of restitution between the spheres is 0.4 and between Q and the wall is e, show that P and Q collide again if $e > \dfrac{1}{7}$

b If $e = \dfrac{1}{10}$, find the total loss of kinetic energy due to the collisions.

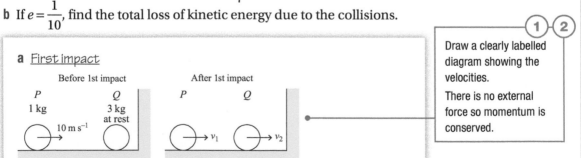

a First impact

Before 1st impact

P 1 kg 10 m s^{-1} Q 3 kg at rest

After 1st impact

P v_1 Q v_2

(1)(2)

Draw a clearly labelled diagram showing the velocities.

There is no external force so momentum is conserved.

(Continued on the next page)

Momentum equation is:

$$10 + 0 = v_1 + 3v_2 \quad \textbf{(1)}$$

③ Write the two equations and solve simultaneously.

Newton's equation is:

$$\frac{v_2 - v_1}{10} = 0.4 \quad \Rightarrow \quad v_2 - v_1 = 4 \quad \textbf{(2)}$$

(1) and **(2)** give $v_1 = -0.5$ and $v_2 = 3.5$

P is now moving away from the wall and Q towards the wall.

<u>Second impact</u>

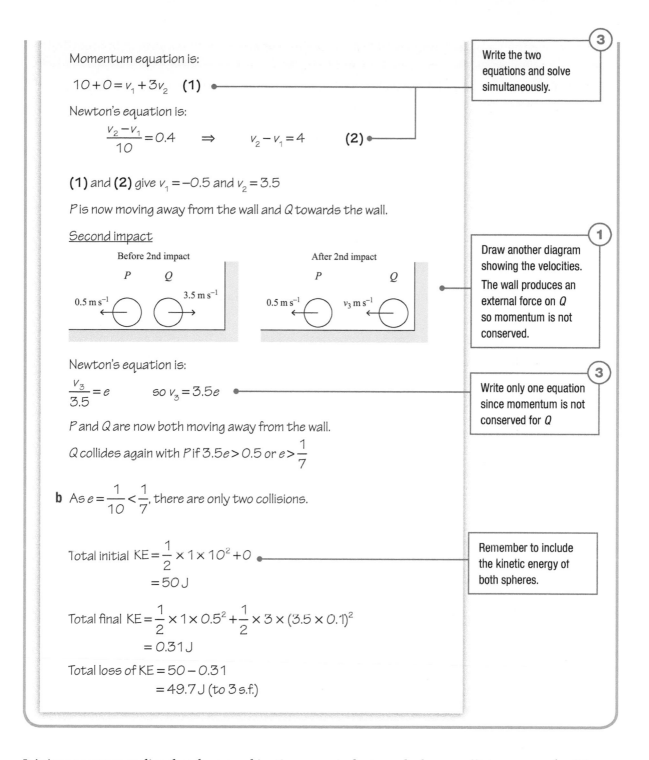

Before 2nd impact

P Q

$0.5\,\text{m s}^{-1}$ $3.5\,\text{m s}^{-1}$

After 2nd impact

P Q

$0.5\,\text{m s}^{-1}$ $v_3\,\text{m s}^{-1}$

① Draw another diagram showing the velocities. The wall produces an external force on Q so momentum is not conserved.

Newton's equation is:

$$\frac{v_3}{3.5} = e \qquad \text{so } v_3 = 3.5e$$

③ Write only one equation since momentum is not conserved for Q

P and Q are now both moving away from the wall.

Q collides again with P if $3.5e > 0.5$ or $e > \dfrac{1}{7}$

b   As $e = \dfrac{1}{10} < \dfrac{1}{7}$, there are only two collisions.

Total initial $KE = \dfrac{1}{2} \times 1 \times 10^2 + 0$

$\qquad\qquad = 50\,\text{J}$

Remember to include the kinetic energy of both spheres.

Total final $KE = \dfrac{1}{2} \times 1 \times 0.5^2 + \dfrac{1}{2} \times 3 \times (3.5 \times 0.1)^2$

$\qquad\qquad = 0.31\,\text{J}$

Total loss of $KE = 50 - 0.31$

$\qquad\qquad\quad = 49.7\,\text{J (to 3 s.f.)}$

It is important to realise that, because kinetic energy is destroyed when a collision occurs (as it is converted into sound energy and heat), the total kinetic energy before the collision is not equal to the total kinetic energy after the collision. Instead, it is momentum which is conserved, as this example shows.

1 A ball of mass m kg is dropped vertically onto a stone floor from a height of 4 metres and rebounds to a height of 3 metres. Using $g = 9.8\,\mathrm{m\,s^{-1}}$, calculate

   a The speed at which it hits the floor and the speed at which it leaves the floor,

   b The coefficient of restitution, e, between the ball and the floor,

   c The loss of kinetic energy due to the impact.

   d Explain why, if the ball is made from wet clay so that $e = 0$, the percentage loss of KE on impact is 100%.

2 Two spheres, P of mass 2 kg and Q of mass 4 kg, both of equal radius, lie on a smooth horizontal plane. P is at rest.

   Q moves away from P at $10\,\mathrm{m\,s^{-1}}$ towards a vertical wall where $e = \dfrac{1}{2}$. Q hits the wall and rebounds to hit and coalesce with P

   a Calculate the final speed of P and Q

   b Calculate the percentage of the total initial KE which is lost.

3 Two balls Y and Z, of equal radius and of mass 4 kg and m kg respectively, lie on a smooth, horizontal surface between two parallel walls, on a line at right angles to them. Y moves at $2\,\mathrm{m\,s^{-1}}$ towards one wall and Z moves at $4\,\mathrm{m\,s^{-1}}$ towards the other wall. They bounce off the walls, where $e = \dfrac{1}{3}$, then collide with each other on their return and bring themselves to rest.

   a Calculate the value of m

   b Which of the two balls loses the least KE in their collisions with the two walls?

4 Two spheres R and S, of equal radius and of mass 2 kg and m kg respectively, lie on a horizontal plane. S moves with speed u away from R (which is at rest) towards a wall fixed at right-angles to its path, rebounds from the wall and returns towards R. The coefficient of restitution for all impacts is 0.75

   a Calculate

     i The value of m such that S is at rest after its impact on R,

     ii The percentage of the total initial KE that is destroyed by both collisions.

   b What assumptions have you made in your solution to this problem?

5 Two spheres, A of mass 1 kg and B of mass 3 kg, both of equal radius, are joined by a light slack string on a smooth horizontal plane. A is at rest. B moves away from A with speed u to strike and rebound from a wall at right angles to its path, with the string still slack. Given that $e = 0.5$ for all collisions, find in terms of u

    a The velocities of A and B after they collide,

    b The common speed of A and B after the string becomes taut.

6 Sphere B of mass 2 kg lies between sphere A of mass 30 kg and a fixed vertical wall, all on a smooth horizontal plane. Both spheres have an equal radius. A is at rest and B moves towards A with speed u. The value of e is 0.6 for all collisions. Show that B is at rest after its second collision with A

7 Three particles P, Q and R, with masses 6 kg, 4 kg and 8 kg respectively, lie in a straight line on a horizontal surface with Q and R at rest. P moves towards Q with velocity 15 m s^{-1}. If $e = \dfrac{3}{5}$ for all collisions, find their velocities after the third collision. Explain why there are no more collisions.

8 A rubber ball of mass m is dropped 5 metres from rest onto a horizontal floor. Taking $g = 10$ m s^{-2}, show that the ball's speed just before the first impact is 10 m s^{-1}. Given $e = 0.5$, calculate the time between the first and second impacts. Use a geometric progression to find the total time for the ball to come to rest after being released.

9 Three particles X, Y and Z, of masses 2 kg, 4 kg and 4 kg respectively, lie in a straight line on a smooth horizontal plane. X and Y are joined by a light slack string. With X and Z at rest, Y moves towards Z at 6 m s^{-1} and the string becomes taut before Y reaches Z. If $e = \dfrac{1}{2}$ for all impacts, calculate the common speed of X and Y

    a Before Y strikes Z,

    b After the string becomes taut for a second time.

Explain why there are no more collisions.

Fluency and skills

You already know that the impulse I of a constant force F acting on a body for a time t is equal to the change in momentum of the body.

Key point

Impulse, $I = F \times t$

$\qquad = mv - mu$

> I, F, v and u are all vectors. You may see J used rather than I for impulse.

F can be very large and t very small, as in a collision.

F can be constant over time, as when a jet of water hits a fixed surface.

F can vary with time, in which case you need to use calculus.

Key point

$$I = \int_0^T F \, dt = \int_0^T m \frac{dv}{dt} \, dt = [mv]_u^v = mv - mu$$

> $F = m \dfrac{dv}{dt}$ comes from Newton's 2nd law of motion.

In each case, impulse = change in momentum.

Example 1

A ball of mass 0.2 kg moves with velocity 8 m s^{-1} horizontally towards a bat. It is struck by the bat, reverses its direction and moves away at 12 m s^{-1}. Find the impulse of the bat on the ball. What is the impulse of the ball on the bat?

Before   During   After

 8 m s^{-1} I 12 m s^{-1}

Impulse equation for the ball alone is

$I = mv - mu$

$I = 0.2 \times 12 - 0.2 \times (-8)$

$\quad = 2.4 + 1.6 = 4.0 \, Ns$

The impulse of the bat on the ball is 4 N s.

From Newton's 3rd law of motion, the impulse of the ball on the bat is 4 N s in the opposite direction.

> Apply the impulse equation for the ball alone. The positive direction is to the right in the diagram.

Example 2

Water flows horizontally from a pipe of cross-section $0.02\,\text{m}^2$ at a speed of $15\,\text{m s}^{-1}$

It strikes a fixed vertical wall.

Find the force F on the wall. (Density of water $= 1000\,\text{kg m}^{-3}$)

Mass of water from pipe in 1 second is

$0.02 \times 15 \times 1000 = 300\,\text{kg}$ — Calculate the mass of water.

Impulse equation for water is

$I = Ft = mv - mu$

$F \times 1 = 0 - 300 \times 15$

Force F on the water $= -4500\,\text{N}$

So, force on the wall $= 4500\,\text{N}$

Calculate the mass of water.

Use the impulse equation to find the constant force F

All the water's momentum is destroyed on impact.

Newton's 3rd law states that forces are equal and opposite.

Example 3

A particle of mass $2\,\text{kg}$ has an initial velocity u of $3\,\text{m s}^{-1}$

It is subject to a force $F = 8t - 3t^2$ newtons for 3 seconds where $0 \leq t \leq 3$
Find its speed v after 3 seconds.

Impulse $I = \int_0^3 F\,dt = \int_0^3 8t - 3t^2\,dt$ — Use calculus.

$= \left[4t^2 - t^3\right]_0^3 = 36 - 27 - 0 = 9\,\text{N s}$

Impulse equation is

$I = mv - mu$ — Solve to find v

$9 = 2v - 2 \times 3$

Final velocity, $v = 7.5\,\text{m s}^{-1}$

1   A ball of mass m kg falls vertically and hits a horizontal floor with a speed u m s^{-1}. It rebounds with speed v m s^{-1}. Find the impulse of the floor on the ball when

   a $m = 2, u = 12, v = 8$

   b $m = 6, u = 25, v = 10$

2   Two identical particles A and B of mass m kg collide head-on. Particle A has an initial velocity u m s^{-1} and a final velocity v m s^{-1}. Find the impulse that A exerts on B when

   a $m = 3, u = 8, v = 2$

   b $m = 5, u = 6, v = -3$

3   **a**   A golf club hits a ball at rest with a horizontal impulse of 48 N s. If the ball has a mass of 50 grams, find its initial velocity.

   b   A cricket bat strikes a ball of mass 0.15 kg horizontally with an impulse of 36 N s. If the ball is moving horizontally towards the bat with a speed 8 m s^{-1}, what is its final velocity?

4   Water flows from a circular pipe of radius r with a speed v and strikes a fixed wall at right angles without any rebound. Given the density of water is 1000 kg m^{-3}, find the force that the water exerts on the wall, when

   a $r = 2$ cm, $v = 12$ m s^{-1}

   b $r = 15$ mm, $v = 20$ m s^{-1}

5   Find the force on each square metre of ground when 1.2 cm of rain falls in 2 hours, striking the ground vertically with a speed of 72 m s^{-1}. (1 m^3 of water has a mass of 1000 kg.)

6   A 5 kg particle with an initial velocity of 2 m s^{-1} is acted on by a variable force F (in N s) for 4 seconds. Find the impulse I acting on the particle and its final velocity, v, when

   a $F = 20t - 6t^2$ **b** $F = 3t^2 - 8t + 1$

Reasoning and problem-solving

Strategy

To solve problems involving impulses,

(1) Draw and label a diagram to show the information.

(2) Consider the situation and briefly explain your strategy and method.

(3) Write and solve equations to calculate the required values.

Example 4

Two masses, A of $2\,kg$ and B of $3\,kg$, both of equal radius, are connected by a slack, inextensible string. Mass B is projected with a velocity of $10\,m\,s^{-1}$ directly away from mass A. Find

a Their final velocities after the string becomes taut,

b The instantaneous impulsive tension in the string.

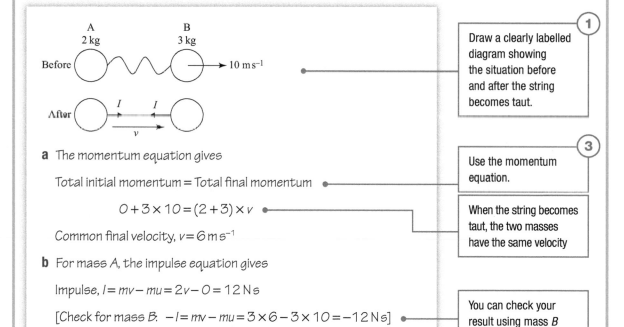

a The momentum equation gives

Total initial momentum = Total final momentum

$$0 + 3 \times 10 = (2 + 3) \times v$$

Common final velocity, $v = 6\,m\,s^{-1}$

b For mass A, the impulse equation gives

Impulse, $I = mv - mu = 2v - 0 = 12\,N\,s$

[Check for mass B: $-I = mv - mu = 3 \times 6 - 3 \times 10 = -12\,N\,s$]

Draw a clearly labelled diagram showing the situation before and after the string becomes taut.

Use the momentum equation.

When the string becomes taut, the two masses have the same velocity

You can check your result using mass B

Example 5

A 2 kg toy plane flying at $6\,\mathrm{m\,s^{-1}}$ in a straight line is subjected to a variable force for 9 seconds, as shown in this graph.

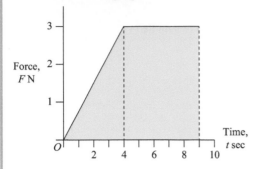

Calculate the impulse of the force and the plane's final velocity.

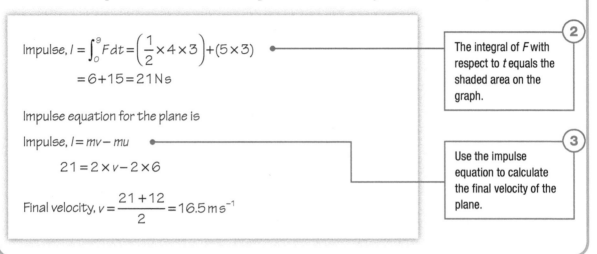

Impulse, $I = \int_0^9 F\,dt = \left(\dfrac{1}{2}\times 4 \times 3\right) + (5 \times 3)$

$= 6 + 15 = 21\,\mathrm{Ns}$

2 The integral of F with respect to t equals the shaded area on the graph.

Impulse equation for the plane is

Impulse, $I = mv - mu$

$21 = 2 \times v - 2 \times 6$

Final velocity, $v = \dfrac{21 + 12}{2} = 16.5\,\mathrm{m\,s^{-1}}$

3 Use the impulse equation to calculate the final velocity of the plane.

Exercise 2.3B Reasoning and problem-solving

1 Particle A of mass 2 kg moving at $12\,\mathrm{m\,s^{-1}}$ collides directly with particle B of mass 3 kg and speed $6\,\mathrm{m\,s^{-1}}$ moving in the same direction. They coalesce. Calculate the impulse of A on B

2 A light, inextensible string connects two masses P and Q with masses of 4 kg and 6 kg respectively. The string is slack when Q is projected with a velocity of $2\,\mathrm{m\,s^{-1}}$ directly away from P. Find the final velocities of the two masses and the instantaneous impulsive tension in the string when it tightens.

3 Three particles A, B and C, of mass 1 kg, 2 kg and 3 kg respectively, are in line and joined by two slack strings AB and BC. After C is projected directly away from B with a velocity of $3\,\mathrm{m\,s^{-1}}$, find the impulsive tension which occurs in each string as it becomes taut.

4 A 2 kg particle has a velocity of $5\,\mathrm{m\,s^{-1}}$ when it is subjected to a variable force F for 6 seconds where $F = \frac{1}{2}t(8-t)$ newtons.
Find the impulse on the particle over this period and also its final velocity.

5 An object of mass 4 kg with an initial velocity of $10\,\mathrm{m\,s^{-1}}$ is acted on by a variable force F (in N) for 12 seconds. Find the impulse acting on the particle over this time and its final velocity, if

$$F = \begin{cases} \frac{1}{2}t^2 & \text{for } 0 \le t \le 4 \\ 12 - t & \text{for } 4 \le t \le 12 \end{cases}$$

6 A truck of mass 400 kg moving at $5\,\mathrm{m\,s^{-1}}$ on a smooth track experiences the variable force shown by this graph. Calculate the impulse of this force on the truck and the truck's final velocity.

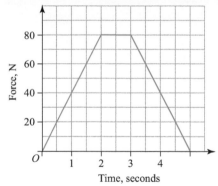

7 A box of mass 20 kg is pushed from rest over a smooth horizontal surface by the variable force shown in the diagram.

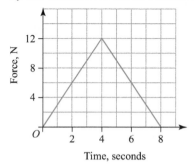

Calculate

a The impulse on the box,

b The box's final velocity,

c The time when its velocity is $0.6\,\mathrm{m\,s^{-1}}$

8 A bullet of mass m and speed u strikes and embeds in a wooden disc of mass M moving in the same direction with speed v. Find the impulse of the bullet on the disc.

9 Two smooth spheres P and Q, of mass 3 kg and 4 kg, both of equal radius, on a smooth table are joined by a slack inelastic string. Q moves at $21\,\mathrm{m\,s^{-1}}$ directly away from P which is at rest.

a At the moment the string becomes taut, find the velocity of both spheres and the impulsive tension in the string.

b Describe the changes in energy that occur during this problem.

10 A shell of mass 1.25 kg is fired horizontally with a speed of $380\,\mathrm{m\,s^{-1}}$ from a gun of mass 50 kg. The gun is brought to rest by a constant horizontal force acting over a distance of 1 metre. Calculate

a The impulse of the explosion and the initial speed of recoil of the gun,

b The magnitude of the force.

Fluency and skills

Full A Level

Key point

Momentum $= m \times v$ Impulse $= F \times t$

Impulse $=$ Change in momentum

$$I = mv - mu$$

In the first three sections of this chapter, you have used these momentum and impulse equations and the principle of conservation of momentum to solve problems where motion has been in one direction. You will now use these equations to solve problems when the motion is in two dimensions.

One way you can do this is to write a vector equation.

Example 1

A particle P of mass 1 kg with a velocity $(2\mathbf{i} - 3\mathbf{j})$ m s^{-1} collides and coalesces with a particle Q of mass 4 kg and velocity $(\mathbf{i} + 2\mathbf{j})$ m s^{-1}. They move off together with a common velocity $\mathbf{v}$. Calculate the speed $|\mathbf{v}|$ and the angle which $\mathbf{v}$ makes with the x-axis.

Final mass $= 1 + 4 = 5$ kg

The momentum equation is

$$1 \times (2\mathbf{i} - 3\mathbf{j}) + 4 \times (\mathbf{i} + 2\mathbf{j}) = 5 \times v$$

$$6\mathbf{i} + 5\mathbf{j} = 5v$$

Final velocity, $v = (1.2\mathbf{i} + 1\mathbf{j})$ m s^{-1}

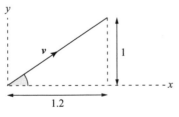

Speed, $|v| = \sqrt{1.2^2 + 1^2} = \sqrt{2.44} = 1.56$ m s^{-1}

at an angle of $\tan^{-1}\dfrac{1}{1.2} = 39.8°$ with the x-axis.

Calculate the combined mass after impact.

There are no external forces, so momentum is conserved.

Calculate the final velocity.

Use Pythagoras and trigonometry to calculate the speed and angle.

You can also solve problems in two dimensions by considering the components separately. For example, when solving a problem involving an impact with a plane, you would solve separate equations for motion parallel to and perpendicular to the plane.

The particle in this diagram strikes a smooth, fixed plane with a velocity $\boldsymbol{u}$ and rebounds with velocity $\boldsymbol{v}$

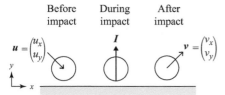

As the plane is smooth, the impulse on impact, $\boldsymbol{I}$, is perpendicular to the plane and momentum in this direction is *not* conserved. There is no impulse parallel to the plane and therefore there is no change in the particle's momentum parallel to the plane, so $u_x = v_x$

However, Newton's law of restitution still applies perpendicular to the plane, so $e = \dfrac{v_y}{u_y}$

Example 2

A 2 kg ball with velocity $\boldsymbol{u} = 3\boldsymbol{i} - 4\boldsymbol{j}$ impacts a smooth plane, which is parallel to the x-axis. Given $e = 0.5$, calculate the final velocity $\boldsymbol{v} = v_x\boldsymbol{i} + v_y\boldsymbol{j}$ of the ball and the impulse $\boldsymbol{I}$ of the plane on the ball.

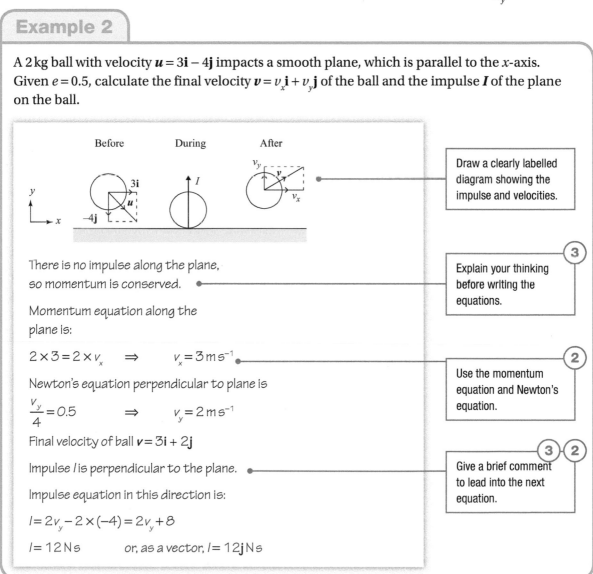

Draw a clearly labelled diagram showing the impulse and velocities.

There is no impulse along the plane, so momentum is conserved.

(3) Explain your thinking before writing the equations.

Momentum equation along the plane is:

$2 \times 3 = 2 \times v_x \quad \Rightarrow \quad v_x = 3\,\mathrm{m\,s^{-1}}$

(2) Use the momentum equation and Newton's equation.

Newton's equation perpendicular to plane is

$\dfrac{v_y}{4} = 0.5 \quad \Rightarrow \quad v_y = 2\,\mathrm{m\,s^{-1}}$

Final velocity of ball $\boldsymbol{v} = 3\boldsymbol{i} + 2\boldsymbol{j}$

Impulse $\boldsymbol{I}$ is perpendicular to the plane.

(3)(2) Give a brief comment to lead into the next equation.

Impulse equation in this direction is:

$I = 2v_y - 2 \times (-4) = 2v_y + 8$

$I = 12\,\mathrm{N\,s} \qquad$ or, as a vector, $I = 12\boldsymbol{j}\,\mathrm{N\,s}$

Example 3 shows how an external impulse creates an internal impulsive tension in a string and causes changes to velocities.

Example 3

Two stationary spheres A and B, of masses $4\,\text{kg}$ and $2\,\text{kg}$ respectively, both of equal radius, are connected by an inelastic string on a smooth horizontal plane. B is struck by an impulse $\boldsymbol{J} = (24\mathbf{i} + 30\mathbf{j})\,\text{N}\,\text{s}$.

a Show that B begins to move at $75°$ to the line AB

b Calculate the impulsive tension in the string, $\boldsymbol{I}$

a

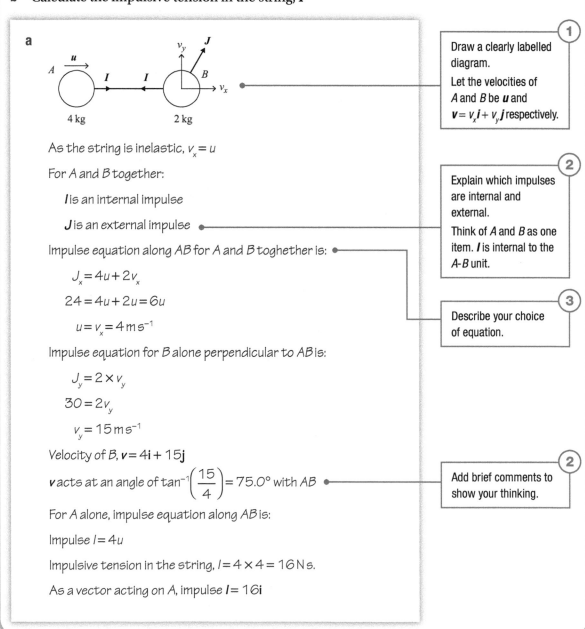

As the string is inelastic, $v_x = u$

For A and B together:

  I is an internal impulse

  J is an external impulse

Impulse equation along AB for A and B toghether is:

  $J_x = 4u + 2v_x$

  $24 = 4u + 2u = 6u$

  $u = v_x = 4\,\text{m}\,\text{s}^{-1}$

Impulse equation for B alone perpendicular to AB is:

  $J_y = 2 \times v_y$

  $30 = 2v_y$

  $v_y = 15\,\text{m}\,\text{s}^{-1}$

Velocity of B, $\boldsymbol{v} = 4\mathbf{i} + 15\mathbf{j}$

$\boldsymbol{v}$ acts at an angle of $\tan^{-1}\left(\dfrac{15}{4}\right) = 75.0°$ with AB

For A alone, impulse equation along AB is:

Impulse $I = 4u$

Impulsive tension in the string, $I = 4 \times 4 = 16\,\text{N}\,\text{s}$.

As a vector acting on A, impulse $\boldsymbol{I} = 16\mathbf{i}$

① Draw a clearly labelled diagram.
Let the velocities of A and B be $\boldsymbol{u}$ and $\boldsymbol{v} = v_x\mathbf{i} + v_y\mathbf{j}$ respectively.

② Explain which impulses are internal and external.
Think of A and B as one item. I is internal to the A–B unit.

③ Describe your choice of equation.

② Add brief comments to show your thinking.

You could check the answer to Example 3 by considering the impulse equation for B alone in the direction AB:

$J_x - I = 2 \times v_x$ giving $I = 24 - 8 = 16$ N s in the direction towards A as in the diagram.

As a vector acting on B, impulse $\boldsymbol{I} = -16\boldsymbol{i}$

Exercise 2.4A Fluency and skills

Full A Level

1   A particle with a velocity $(4\boldsymbol{i} - 6\boldsymbol{j})\,\mathrm{m\,s^{-1}}$ collides and coalesces with a similar particle with velocity $(\boldsymbol{i} + 2\boldsymbol{j})\,\mathrm{m\,s^{-1}}$. Both particles have mass 500 grams. Calculate their velocity, v, after impact.

2   Particle P of mass 2 kg moves with velocity $(3\boldsymbol{i} + 3\boldsymbol{j})\,\mathrm{m\,s^{-1}}$ and collides with particle Q of mass 3 kg moving with velocity $(4\boldsymbol{i} - 1\boldsymbol{j})\,\mathrm{m\,s^{-1}}$. After impact, P is at rest. Calculate the velocity of Q

3   Particle P has mass 2 kg and velocity $(5\boldsymbol{i} - \boldsymbol{j})\,\mathrm{m\,s^{-1}}$. Particle Q has mass 4 kg and velocity $(-\boldsymbol{i} - 4\boldsymbol{j})\,\mathrm{m\,s^{-1}}$. They collide and coalesce. Calculate their velocity after impact.

4   Particle S of mass 4 kg and velocity $(5\boldsymbol{i} + 4\boldsymbol{j})\,\mathrm{m\,s^{-1}}$ collides with particle T of mass 2 kg and velocity $(-2\boldsymbol{i} + \boldsymbol{j})\,\mathrm{m\,s^{-1}}$. After impact, they move off in the same straight line. T moves twice as fast as S. Calculate their final velocities.

5   At a fireworks display, a rocket of mass 0.5 kg travels in a vertical plane. When its velocity is $(6\boldsymbol{i} + 20\boldsymbol{j})\,\mathrm{m\,s^{-1}}$, it splits into two pieces with masses 0.3 kg and 0.2 kg with velocities $(-2\boldsymbol{i} + 18\boldsymbol{j})\,\mathrm{m\,s^{-1}}$ and $(a\boldsymbol{i} + b\boldsymbol{j})\,\mathrm{m\,s^{-1}}$ respectively. Calculate a and b

6   A sphere strikes a smooth horizontal surface containing the x-axis with a velocity $\boldsymbol{u} = (u_x\boldsymbol{i} - u_y\boldsymbol{j})\,\mathrm{m\,s^{-1}}$. It rebounds with a velocity $\boldsymbol{v} = (v_x\boldsymbol{i} + v_y\boldsymbol{j})\,\mathrm{m\,s^{-1}}$. The coefficient of restitution is e

   a   Copy and complete this table for the four spheres A to D

	u_x	u_y	e	v_x	v_y
A	4	3	$\frac{1}{2}$		
B	5	2	$\frac{1}{4}$		
C	3	4			3
D		2		4	1

   b   In each case, find the magnitude of the final velocity $\boldsymbol{v}$ and the angle it makes with the surface just after impact.

7   A particle moving with velocity $(3\boldsymbol{i} - 4\boldsymbol{j})\,\mathrm{m\,s^{-1}}$ strikes a smooth horizontal surface. If the coefficient of restitution is 0.5, find its velocity $\boldsymbol{v} = (v_x\boldsymbol{i} + v_y\boldsymbol{j})\,\mathrm{m\,s^{-1}}$ after impact. Also calculate the angle between the old direction and the new direction of the particle.

8 A ball moves with velocity $(5\mathbf{i} - 3\mathbf{j})\,\mathrm{m\,s^{-1}}$ towards a smooth plane which contains the x-axis. The coefficient of restitution e is 0.8. Calculate the speed of the ball immediately after impact and the angle it makes with the plane.

9 A 5 kg ball with velocity $\boldsymbol{u} = (2\mathbf{i} - 3\mathbf{j})\,\mathrm{m\,s^{-1}}$ strikes a smooth plane containing the x-axis. If $e = 0.2$, calculate the final velocity $\boldsymbol{v} = (v_x\mathbf{i} + v_y\mathbf{j})\,\mathrm{m\,s^{-1}}$ of the ball and the impulse $\boldsymbol{J}$ of the plane on the ball.

10 A sphere with velocity $\boldsymbol{u} = (-4\mathbf{i} + 3\mathbf{j})\,\mathrm{m\,s^{-1}}$ impacts a smooth plane containing the y-axis with $e = 0.6$. If the sphere's mass is 6 kg, calculate its velocity $\boldsymbol{v} = (v_x\mathbf{i} + v_y\mathbf{j})\,\mathrm{m\,s^{-1}}$ after impact and the impulse $\boldsymbol{J}$ of the plane on the sphere.

11 A small 2 kg ball, moving in the horizontal x-y plane with a velocity $\boldsymbol{u} = (-16\mathbf{i} - 12\mathbf{j})\,\mathrm{m\,s^{-1}}$, strikes a smooth wall containing the x-axis. It subsequently strikes another wall containing the y-axis. If $e = \dfrac{3}{4}$ for both collisions, find

a Its final velocity as a vector just after the second collision,

b The total loss of kinetic energy due to the two collisions.

Reasoning and problem-solving

Full A Level

When two spheres moving in the same straight line collide, they undergo a *direct* collision. However, when two spheres collide from different directions, an **oblique impact** occurs.

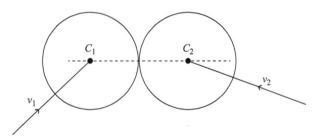

During an oblique impact of two spheres, there is an internal impulse, $\boldsymbol{I}$, along the line of centres C_1C_2 but no *external* impulse, so momentum is conserved. You can write a momentum equation and use Newton's law in this direction.

There is no impulse perpendicular to C_1C_2, so the components of each velocity in this direction are unchanged and momentum is conserved in this direction too.

> To solve problems in two dimensions, you may have to resolve vectors to find the component in each direction.

To solve problems involving momentum and impulse in two dimensions:

1. Draw and label a diagram clearly showing the situation at different stages of the motion.

2. Check whether there are any external forces and whether momentum is conserved.

3. Write and solve equations using impulse, momentum and Newton's law of restitution.

Example 4

Sphere P of mass $4\,\text{kg}$ and speed $8\,\text{m s}^{-1}$ strikes sphere Q of mass $2\,\text{kg}$ and speed $4\,\text{m s}^{-1}$. Their speeds before impact are at $30°$ and $60°$ to the line of their centres C_1C_2.

Given that $e = \dfrac{1}{2}$, find the magnitude and direction of their velocities after impact. Assume the spheres have an equal radius.

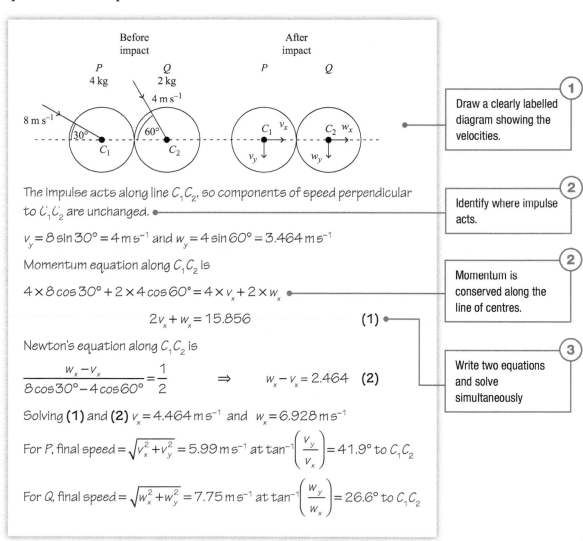

The impulse acts along line C_1C_2, so components of speed perpendicular to C_1C_2 are unchanged.

1. Draw a clearly labelled diagram showing the velocities.

2. Identify where impulse acts.

$v_y = 8\sin 30° = 4\,\text{m s}^{-1}$ and $w_y = 4\sin 60° = 3.464\,\text{m s}^{-1}$

Momentum equation along C_1C_2 is

$4 \times 8\cos 30° + 2 \times 4\cos 60° = 4 \times v_x + 2 \times w_x$

2. Momentum is conserved along the line of centres.

$$2v_x + w_x = 15.856 \qquad (1)$$

Newton's equation along C_1C_2 is

$$\frac{w_x - v_x}{8\cos 30° - 4\cos 60°} = \frac{1}{2} \qquad \Rightarrow \qquad w_x - v_x = 2.464 \quad (2)$$

3. Write two equations and solve simultaneously

Solving (1) and (2) $v_x = 4.464\,\text{m s}^{-1}$ and $w_x = 6.928\,\text{m s}^{-1}$

For P, final speed $= \sqrt{v_x^2 + v_y^2} = 5.99\,\text{m s}^{-1}$ at $\tan^{-1}\left(\dfrac{v_y}{v_x}\right) = 41.9°$ to C_1C_2

For Q, final speed $= \sqrt{w_x^2 + w_y^2} = 7.75\,\text{m s}^{-1}$ at $\tan^{-1}\left(\dfrac{w_y}{w_x}\right) = 26.6°$ to C_1C_2

1 A particle of mass $3\,\text{kg}$ has a velocity of $4\,\text{m}\,\text{s}^{-1}$ in the x-direction. A second particle has mass $m\,\text{kg}$ and velocity $6\,\text{m}\,\text{s}^{-1}$ in the y-direction. They collide, coalesce and begin to move together at $45°$ to their initial directions. Calculate the magnitude of their final velocity.

2 Sphere A of mass $4\,\text{kg}$ and centre C_1 moves at $6\,\text{m}\,\text{s}^{-1}$ at $60°$ to the x-axis. Sphere B of mass $2\,\text{kg}$ is at rest with centre C_2 on the x-axis. The spheres are of equal radius. A strikes B so that the line of centres C_1C_2 is the x-axis. After impact, A moves at right angles to the x-axis. Explain why B moves along the x-axis and calculate the final velocities of A and B. The spheres are on a smooth horizontal surface.

3 Two smooth spheres, of mass $4\,\text{kg}$ and $8\,\text{kg}$ and of equal radius, collide obliquely. The first has a velocity of $10\,\text{m}\,\text{s}^{-1}$ at $60°$ to their line of centres. The second is at rest. If $e = 0.5$, find the magnitudes and directions of their velocities just after impact.

4 Two smooth spheres, of mass $4\,\text{kg}$ and $2\,\text{kg}$ and of equal radius, move with speeds of $8\,\text{m}\,\text{s}^{-1}$ and $4\,\text{m}\,\text{s}^{-1}$ at angles of $30°$ and $60°$ respectively to their line of centres. They collide with $e = 0.5$. Find their speeds immediately after impact and the loss of kinetic energy due to the impact.

5 A collision occurs between two similar balls moving on a horizontal surface with velocities of $(5\mathbf{i} + 6\mathbf{j})\,\text{m}\,\text{s}^{-1}$ and $(-2\mathbf{i} + 4\mathbf{j})\,\text{m}\,\text{s}^{-1}$. On impact, their line of centres is parallel to the direction $\mathbf{i}$ and, after impact, their velocities are $(\mathbf{i} + 6\mathbf{j})\,\text{m}\,\text{s}^{-1}$ and $(2\mathbf{i} + 4\mathbf{j})\,\text{m}\,\text{s}^{-1}$ respectively.

  a Find the coefficient of restitution, e

  b What assumptions have you made in the model used in your solution?

6 A smooth $4\,\text{kg}$ ball with a speed of $4\,\text{m}\,\text{s}^{-1}$ collides with a smooth $2\,\text{kg}$ ball at rest. Both balls are of equal radius. On impact, the direction of the moving ball is at $60°$ to their line of centres. If $e = \dfrac{1}{3}$, find the loss of KE due to the collision and the impulse between the two balls during impact.

7 An $8\,\text{kg}$ ball with a speed of $4\,\text{m}\,\text{s}^{-1}$ collides with a $4\,\text{kg}$ ball with a speed of $2\,\text{m}\,\text{s}^{-1}$. Both balls are of equal radius. Just before impact, their velocities make angles of $30°$ and $60°$ respectively with their line of centres. Given that $e = \dfrac{1}{2}$, find their velocities after impact, the loss of KE due to impact and the impulse between them during impact.

8 Two spheres of equal radius collide with their line of centres parallel to the x-axis. Before impact, sphere A of mass $4\,\text{kg}$ has a velocity $\boldsymbol{u} = (2\mathbf{i} + 3\mathbf{j})\,\text{m}\,\text{s}^{-1}$ and sphere B of mass $2\,\text{kg}$ is at rest.

  a Explain why B moves off in the direction of the x-axis after impact.

  b If $e = 0.5$, find their velocities after impact and the impulse between them during impact.

9 When two spheres of equal radius, P of mass $2\,\text{kg}$ and Q of mass $4\,\text{kg}$, collide, their line of centres is parallel to the y-axis. Before impact, the velocity of P is $\boldsymbol{u} = 4\mathbf{i} + 6\mathbf{j}\,\text{m}\,\text{s}^{-1}$ and Q is at rest. Given $e = 0.4$, find their velocities immediately after impact and the impulse between them during impact.

10 Two spherical objects of the same size but with masses of 6 kg and 4 kg move on a smooth horizontal table with velocities of $(4\mathbf{i} + 2\mathbf{j})\,\mathrm{m\,s^{-1}}$ and $(-2\mathbf{i} + \mathbf{j})\,\mathrm{m\,s^{-1}}$ respectively. They collide when their line of centres is parallel to the x-axis. If $e = 0.25$,

    a Find their velocities as vectors just after the collision and the angle between their directions,

    b Find the impulsive force between them during impact.

11 A ball of mass 2 kg with velocity $(4\sqrt{3}\mathbf{i} + 4\mathbf{j})\,\mathrm{m\,s^{-1}}$ strikes a second ball of mass 4 kg with velocity $(\sqrt{3}\mathbf{i} + \mathbf{j})\,\mathrm{m\,s^{-1}}$. Both balls are of equal radius. Just before impact, they are moving in parallel directions making $30°$ with the line of their centres. Given $e = \dfrac{1}{3}$, calculate their velocities after impact and the impulse between them during impact.

12 A ball of mass 2 kg with speed $10\,\mathrm{m\,s^{-1}}$ strikes a stationary ball of mass 4 kg. Both balls are of equal radius. The direction of the moving ball makes a $60°$ angle with the line of centres at the moment of impact. Given $e = \dfrac{1}{2}$, calculate the velocities of both balls after impact.

13 a A snooker player hits a ball with velocity u at $60°$ to side AB of a snooker table ABCD so that it rebounds from AB and then BC. If $e = \dfrac{1}{\sqrt{3}}$ for both impacts, find its velocity on leaving BC and the total angle it has turned through.

    b The player now hits a ball with velocity u at angle θ to side AB so that the ball rebounds from each of the four sides in turn. Show that after the fourth impact the ball is moving parallel to its initial direction regardless of the value of e

    c If the snooker player strikes the stationary ball 'off-centre', describe, without any calculation, how this affects the motion of the ball. What would a refined model take into account?

14 A ball drops vertically and strikes a slope with a speed of $10\,\mathrm{m\,s^{-1}}$. After impact, it moves horizontally. If the slope makes an angle of $30°$ with the horizontal, calculate the coefficient of restitution.

15 A ball, starting at rest, falls freely 10 metres onto a plane inclined at $45°$ to the horizontal. If $e = 0.2$, calculate the speed and direction of the ball immediately after impact. Use $g = 10\,\mathrm{m\,s^{-2}}$

16 A ball P of mass 6 kg moving at $10\,\mathrm{m\,s^{-1}}$ collides with a ball Q of mass 12 kg at rest. Both balls are of equal radius. The direction of motion of P makes an angle of $30°$ with their line of centres. Given $e = 0.75$, find the velocities of the balls after impact and the impulse during their contact.

17 Two identical spheres of mass m move with the same speed u. They collide when their centres are on the x-axis and their directions of motion are at $30°$ and $60°$ to the x-axis.

    a Find their speeds after impact and the impulse between them during impact if the spheres are perfectly elastic.

    b Describe two limitations to the model adopted in this problem.

18 Two particles P and Q of masses M and m are connected by a light rod of length a and are at rest on a table. P is struck with a blow of impulse $\boldsymbol{I}$ at an angle θ to the rod, which creates an impulsive thrust in the rod.

    a   Show that P initially moves at an angle ϕ to the rod given by $\tan\phi = \dfrac{M+m}{M}\tan\theta$

    b   State two assumptions you have made in solving this problem.

    c   State two improvements that a new model could make on this one.

19 Two smooth spheres, of masses m_1 and m_2 and of equal radius, collide obliquely. Just before impact, the first sphere travels at angle θ to their line of centres and the second sphere is at rest. Show that, if $m_1 = e \times m_2$ where e is the coefficient of restitution, then they move off in directions perpendicular to each other.

20 Two identical smooth spheres of mass m have velocities equal in magnitude u but opposite in direction. They collide obliquely with their line of centres at $45°$ to their directions. Find the loss in KE due to the impact and show that it equals half of the loss which would have occurred had it been a direct collision.

21 Two equal particles A and B of mass m are 1 metre apart on a smooth table and joined by a slack inelastic string 2 metres long. B moves with velocity u at right angles to the initial line AB. Find the impulse in the string and the magnitude of the velocity of B at the instant the string becomes taut if

    a A is fixed,   **b** A is free to move.

22 Two equal masses collide with $e = \dfrac{1}{3}$. Before impact, one mass has a speed of u making an angle θ to the line of centres on impact and the other is at rest. If the direction of the first mass is at an angle φ to the line of centres after impact, show that $\tan\varphi = 3\tan\theta$

23 Two smooth spheres P and Q are equal in radius. P has mass m and Q has mass km where $k > 1$. P collides with Q at rest. P's initial direction of motion is at an angle θ to their line of centres, where $\theta > 0$, and the impact deflects P through $90°$. Prove that the coefficient of restitution $e > \dfrac{1}{k}$. Show that $\tan^2\theta \geq \dfrac{k-1}{k+1}$

Chapter summary

- The momentum of a body is its mass multiplied by its velocity;
  momentum $= m \times \boldsymbol{v}$

- The principle of conservation of momentum applies when no external forces are present.
  It says that the total momentum is constant.
  So, Total initial momentum = Total final momentum.

- When two bodies collide, Newton's law of restitution gives the equation
  $$\frac{\text{Speed of separation}}{\text{Speed of approach}} = e$$

  This can also be written as

  $$e = \frac{v_2 - v_1}{u_1 - u_2}$$

- The constant e is known as the coefficient of restitution.
  If $e = 0$, the impact is inelastic.
  If $e = 1$, the impact is perfectly elastic.
  In general, $0 \leq e \leq 1$

- When a force $\boldsymbol{F}$ acts on a body for a time t, the impulse $\boldsymbol{I} = \boldsymbol{F} \times t$ when $\boldsymbol{F}$ is constant
  or $\boldsymbol{I} = \int \boldsymbol{F} \, dt$ when $\boldsymbol{F}$ is variable.

- An impulse creates a change in momentum. The impulse equation is
  Impulse of force = Change in momentum produced

- Velocity, momentum, force and impulse are all vectors, so equations involving them must take account of their directions.   `Full A Level`

- When two smooth spheres collide obliquely

  - Along the line of centres, momentum is conserved and Newton's law applies,

  - Perpendicular to the line of centres, there is no impulse and so no change in momentum.

Check and review

You should now be able to...	Try Questions
✔ Use the momentum equation and Newton's equation appropriately.	1–16
✔ Find the size of an impulse for constant and variable forces.	4, 6, 12, 13, 16
✔ Apply these concepts to problems in 2 dimensions. `Full A Level`	9–16

1   Particle A of mass 2 kg moves with velocity $10\,\mathrm{m\,s^{-1}}$. It collides directly with another particle B of mass 4 kg moving towards A with velocity $8\,\mathrm{m^{-1}}$. After the collision, B is at rest. Find the final velocity of A

2   A particle with mass 2 kg and speed $20\,\mathrm{m\,s^{-1}}$ collides head-on with a particle with mass 6 kg and speed $6\,\mathrm{m\,s^{-1}}$.

   a   If the coefficient of restitution $e = \dfrac{1}{2}$, calculate their speeds after impact.

   b   What assumptions have you made in the model used in your solution?

3   A ball drops from rest from a height of 9 m onto a horizontal plane and rebounds to a height of 1 m. The ball then continues to bounce.

   a   Calculate the value of e and the total distance it has travelled when it hits the ground for the third time. (Use $g = 10\,\mathrm{m\,s^{-2}}$)

   b   If the ball is made from a substance for which $e = 0$, what can you say about the kinetic energy of the ball immediately after striking the wall?

4   A railway truck of mass 10 tonnes and speed $3\,\mathrm{m\,s^{-1}}$ strikes and couples with a lighter truck of 5 tonnes at rest on the same line. Find their common speed after impact and the impulse in the coupling between them during impact. Note, 1 tonne = 1000 kg

5   Sphere A of mass 3 kg moving at $12\,\mathrm{m\,s^{-1}}$ collides directly with sphere B of mass m kg, moving at $2\,\mathrm{m\,s^{-1}}$ in the same direction. Both spheres have an equal radius. After the impact, A moves at $2\,\mathrm{m\,s^{-1}}$ in the opposite direction and B continues at $6\,\mathrm{m\,s^{-1}}$. Calculate the values of m and e and the size of the impulse during the collision.

6   Water flows horizontally from a pipe of cross-section $0.03\,\mathrm{m^2}$ at a speed of $25\,\mathrm{m\,s^{-1}}$. It strikes a fixed vertical wall. Take the density of water as $1000\,\mathrm{kg\,m^{-3}}$ and find the force F on the wall.

7   A particle of mass 2 kg has an initial velocity of $9\,\mathrm{m\,s^{-1}}$. A force $F = 3t^2 - 6t + 2$ newtons acts on it for 3 seconds. Find the impulse on the particle and its speed after 3 seconds.

8   Three particles A, B and C, with masses of 1 kg, 2 kg and 4 kg respectively, are in the same line in this order with A at rest. B and C collide head-on with velocities of $10\,\mathrm{m\,s^{-1}}$ and $8\,\mathrm{m\,s^{-1}}$ respectively. Given that $e = 0.5$ for this and any subsequent impacts, show that B then collides with A. Find all three velocities after this second collision and explain why there are no more collisions.

9   A sphere of mass 2 kg has a velocity $(5\mathbf{i} + 3\mathbf{j})\,\mathrm{m\,s^{-1}}$. It collides and coalesces with a sphere, of equal radius and of mass 4 kg, with a velocity $(2\mathbf{i} - 3\mathbf{j})\,\mathrm{m\,s^{-1}}$. Their common velocity after the collision is $\mathbf{v}$. Calculate the speed $|\mathbf{v}|$ and the angle which $\mathbf{v}$ makes with the x-axis.

10 A particle of mass $5\,\text{kg}$ has a velocity of $4\,\text{m s}^{-1}$ in the x-direction. A second particle has a mass $m\,\text{kg}$ and a velocity of $2\,\text{m s}^{-1}$ in the y-direction. They collide, coalesce and begin to move together at $45°$ to their initial directions. Find their final velocity and the value of m

11 A smooth wall lies along the x-axis. A $4\,\text{kg}$ ball with velocity $\boldsymbol{u} = (5\boldsymbol{i} - 2\boldsymbol{j})\,\text{m s}^{-1}$ strikes the wall where $e = 0.5$. Calculate the final velocity $\boldsymbol{v} = (v_1\boldsymbol{i} + v_2\boldsymbol{j})\,\text{m s}^{-1}$ of the ball and the impulse $\boldsymbol{I}$ of the wall on the ball. State two assumptions you have made in your solution.

12 Two spheres, P of mass $2\,\text{kg}$ and Q of mass $4\,\text{kg}$, both of equal radius, lie at rest on the x-axis with P nearer the origin. They are connected by a straight string. Q is struck by an impulse $\boldsymbol{I} = (24\boldsymbol{i} + 20\boldsymbol{j})\,\text{N s}$. Find the impulsive tension in the string and the spheres' velocities immediately after the blow.

13 A ball of mass $1\,\text{kg}$ strikes a smooth fixed plane with a velocity of $20\,\text{m s}^{-1}$ at $30°$ to the plane where e is 0.4. Find the angle that the final velocity makes with the plane and also the impulse on the ball during the collision.

14 Two smooth spheres, of mass $8\,\text{kg}$ and $4\,\text{kg}$, both of equal radius, have speeds of $12\,\text{m s}^{-1}$ and $6\,\text{m s}^{-1}$ respectively. They collide at angles of $30°$ and $60°$ respectively to their line of centres. If $e = 0.5$, find the magnitude and direction of their velocities immediately after impact and the loss of kinetic energy due to the impact.

15 Two similar smooth balls P and Q move in perpendicular directions at the same speed, u. P moves along their line of centres just before they collide. Prove that Q is deflected through an angle of $\tan^{-1}\left(\dfrac{1+e}{2}\right)$ where e is the coefficient of restitution.

16 Four identical balls of mass m are joined by four straight strings to make a square with the balls at the corners. One ball is struck by an impulse $\boldsymbol{I}$ along a diagonal of the square in the direction pointing outwards. Find the initial velocities of all four balls in terms of m and $\boldsymbol{I}$. What assumption have you made about the string?

Information

The term momentum refers to a 'quantity of motion of a moving body'. It is borrowed from Latin, where it means 'movement, moving power'. The concept of momentum was first introduced by the French mathematician Descartes, who is perhaps most famous for his development of Cartesian graphs.

Investigation

A Newton's cradle is a device that is named after Sir Isaac Newton, and can be used to demonstrate conservation of momentum and energy. The device consists of a number of spheres (usually 5 or 6) suspended so that, when at rest, their centres are all at the same height. When a sphere at the end is lifted and released, it strikes the stationary spheres – a force is transmitted through the stationary spheres and pushes the last sphere out and upward.

Investigate, using ideas of conservation of momentum and energy, what will happen if first one ball is lifted at the end of the cradle and released. Then think about the case for two balls, three balls, and four balls.

History

Sir Isaac Newton first presented his three laws of motion in the "Principia Mathematica Philosophiae Naturalis" in 1686. His second law defines a **force** to be equal to the rate of change of momentum with respect to time, with momentum being defined to be the product of the mass, m, and its velocity, v

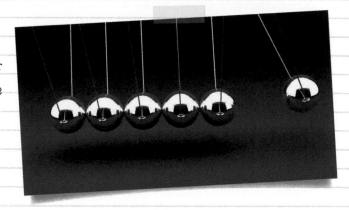

Research

In physics, **angular momentum** is the rotational equivalent of linear momentum. It is an important quantity in mechanics because it is a conserved quantity, and so the total angular momentum of a system remains constant unless acted on by an external **torque**. A torque is the rotational equivalent of a force and provides a twist to an object.

Ideas of angular momentum and torque are important in understanding the motion of a gyroscope. Research gyroscopes and write a report on what they are, how they work, and how they can be used.

1   An object of mass 7 kg, travelling at a speed of $4\,\text{m s}^{-1}$, is acted on by a constant force in its direction of travel which increases its speed to $10\,\text{m s}^{-1}$. Calculate

   a   The impulse exerted on the object,   **[2 marks]**

   b   The force involved if the process took 0.35 seconds.   **[2]**

2   A particle of mass 2 kg is travelling in a straight line at $8\,\text{m s}^{-1}$. A variable braking force acting along the same line of travel is applied so that, after $t\,$s, the magnitude of the force is $2t\,$N. Calculate the time taken for the particle to come to rest.   **[3]**

3   A particle of mass 5 kg is travelling with a velocity of $(4\mathbf{i}+\mathbf{j})\,\text{m s}^{-1}$ when it is subjected to an impulse of $(2\mathbf{i}-7\mathbf{j})\,\text{N s}$. Calculate the new velocity of the particle.   **[3]**

4   A particle of mass 3 kg has velocity $(2\mathbf{i}-3\mathbf{j})\,\text{m s}^{-1}$. It is acted on by a constant force of $(-\mathbf{i}+2\mathbf{j})\,$N, which changes its velocity to $0.5\mathbf{i}\,\text{m s}^{-1}$. For how long does the force act?   **[4]**

5   A snooker ball, moving at $12\,\text{m s}^{-1}$, strikes a smooth cushion at an angle of $25°$. If the coefficient of restitution between the ball and the cushion is 0.7, find the velocity of the ball after the collision.   **[7]**

6   A bullet of mass 0.1 kg is fired horizontally at a block of wood of mass 2 kg, which is stationary and free to move. The bullet enters the block travelling at $100\,\text{m s}^{-1}$. Calculate the subsequent speed of the block if

   a   The bullet passes through the block (but the block loses no mass) and emerges travelling at $40\,\text{m s}^{-1}$,   **[2]**

   b   The bullet becomes embedded in the block.   **[2]**

7   A railway truck of mass $3m$, travelling at a speed of $2v$, collides with another truck, of mass $4m$, travelling with a speed of v. The trucks become coupled together. Find, in terms of v, the common speed of the trucks if, before impact, they were travelling

   a   In the same direction,   **[2]**

   b   In opposite directions.   **[2]**

8   A particle A, of mass 10 kg, is moving at $5\,\text{m s}^{-1}$ when it collides with a particle B, of mass m kg, travelling in the opposite direction at $2\,\text{m s}^{-1}$. After the collision, A travels in the same direction as before but with its speed reduced to $3\,\text{m s}^{-1}$.

   a   If $m=3$, find the velocity of B after the collision.   **[3]**

   b   Show that the value of m cannot be greater than 4   **[4]**

9   Smooth uniform spheres A and B, of equal size, have masses of 3 kg and 2 kg respectively and move on a smooth horizontal surface. A has velocity $(4\mathbf{i}+3\mathbf{j})\,\text{m s}^{-1}$ and B has velocity $(-\mathbf{i}-5\mathbf{j})\,\text{m s}^{-1}$. They collide when the line joining their centres is parallel with $\mathbf{j}$. The coefficient of restitution between the spheres is 0.5

   a   Calculate the velocities of A and B after the collision.   **[5]**

   b   Calculate the impulse received by B **[2]**

   c   Calculate the angle through which the motion of A has been deflected by the collision.   **[2]**

Full A Level

Full A Level

10 A sledgehammer of mass 6 kg, travelling at 20 m s^{-1}, strikes the top of a post of mass 2 kg, which rests in soft ground, and maintains contact with the post.

   a   Calculate the common speed of the hammer and post immediately after impact. **[2]**

   b   The post is brought to rest in 0.02 s by the action of a resistive force R from the ground. By modelling R as constant, find its magnitude. **[2]**

   c   If, in fact, the force is given by $R = k(1 + 2t)$ N, where t s is the time from the moment of impact, find the value of the constant k **[3]**

   d   In what ways would the situation be different if the sledgehammer were to rebound on impact? **[2]**

11 An object of mass 3 kg has velocity $(3\mathbf{i} + 2\mathbf{j})$ m s^{-1}. It collides with another object, which has a mass of 2 kg and a velocity of $(\mathbf{i} - \mathbf{j})$ m s^{-1}. After the impact, the first object has a velocity of $(2\mathbf{i} + \mathbf{j})$ m s^{-1}. Calculate the velocity of the second object. **[2]**

12 Particles A and B have masses of 3 kg and 2 kg respectively. They are connected by a light inextensible string. The particles lie at rest on a smooth horizontal surface. The coefficient of restitution between the particles is 0.5. A is projected towards B with velocity 10 m s^{-1}

   a   Calculate

      i   The velocities of the particles after the collision, **[4]**

      ii   The common velocity of the particles after the string becomes taut. **[2]**

   b   i   Explain why the answer to **a ii** is independent of e, provided $e > 0$ **[1]**

      ii   What would happen if $e = 0$? **[1]**

13 A particle A, with mass 2 kg and velocity 10 m s^{-1}, is moving on a smooth horizontal surface. It catches up and collides with a second particle B, with mass 1 kg and velocity 5 m s^{-1}. B then impacts head-on with a vertical wall. The coefficient of restitution between A and B is 0.5, and between B and the wall is 0.75

   a   Calculate the velocities of A and B after they collide for the first time. **[4]**

   b   Calculate the velocities of the particles after they collide for a second time. **[4]**

   c   Work out what happens after this second impact. **[4]**

14 Three particles A, B and C have masses 3 kg, 2 kg and 1 kg respectively. They are moving, in that order, along a straight line with velocities of 3 m s^{-1}, 2 m s^{-1} and 1 m s^{-1} respectively. The collision between A and B, which happens first, is perfectly elastic. The second collision, between B and C, has coefficient of restitution e

   a   Find the velocities of A and B after the first collision. **[4]**

   b   Find, in terms of e, the velocity of B after the second collision. **[3]**

   c   Hence show that there will be no more collisions if $e \leq \dfrac{4}{11}$ **[3]**

15 The diagram shows a snooker table *ABCD*. The sides *AB* and *AD* are parallel to the **i**- and **j**-directions, as shown.

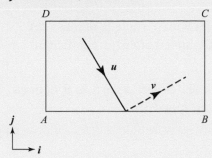

A ball of mass 0.16 kg, travelling with velocity $u = (5\mathbf{i} - 7\mathbf{j})\,\text{m s}^{-1}$, strikes the cushion *AB*, which is assumed to be smooth, and rebounds with velocity $v\,\text{m s}^{-1}$. The coefficient of restitution between the ball and the cushion is 0.7

a **i** Explain why the *x*-component of *v* is $5\,\text{m s}^{-1}$ [1]

   ii Calculate the *y*-component of *v* [2]

   iii Find the angle through which the direction of the ball has been changed. [4]

b The ball is in contact with the cushion for 0.1 s. The force exerted by the cushion has magnitude $kt^2(3 - 8t)\,\text{N}$, where *t* s is the time since first contact. Find the value of the constant *k* [4]

c Do you think that the contact time (0.1 s) stated in part b is realistic? Explain your answer. [2]

16 A ball, *A*, is projected along a smooth surface towards a second, identical, ball, *B*, which is at rest. The speed of *A* is $8\,\text{m s}^{-1}$, at an angle of 30° with the line of centres at the moment of impact. After the collision, *B* has a speed of $4\,\text{m s}^{-1}$. Find

a The coefficient of restitution, [6]

b The velocity of *A* after impact. [4]

17 The diagram shows particles *A*, *B* and *C* of masses 1 kg, 2 kg and 2 kg respectively, which are initially at rest on a straight line. *A* and *B* are connected by a light inextensible string which is slack. *B* is then propelled towards *C* at $6\,\text{m s}^{-1}$. The coefficient of restitution between *B* and *C* is 0.5. The string becomes taut before *B* reaches *C*

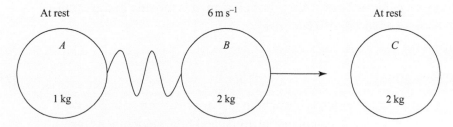

a Find the common speed of *A* and *B* as the string becomes taut. [2]

b Find the impulse on *C* when *B* collides with it. [4]

c Show that, after the string becomes taut for a second time, there are no more collisions. [4]

18 Particles of mass m and $2m$, both travelling with speed u, collide head-on. After the collision, one particle is travelling at twice the speed of the other. Find their speeds after impact in terms of u [3]

19 Particle A of mass m, moving at $7\,\text{m s}^{-1}$, catches up and collides with particle B, of mass km, moving at $1\,\text{m s}^{-1}$. After the collision, the speed of B is twice the speed of A. The coefficient of restitution is 0.75. Find the two possible values of k [5]

20 A particle A, of mass $4\,\text{kg}$, traveling at $6\,\text{m s}^{-1}$ catches up and collides with a particle B, of mass $m\,\text{kg}$, travelling at $1\,\text{m s}^{-1}$. The particles coalesce and move with speed $3\,\text{m s}^{-1}$

   a Calculate the value of m [2]

   b What would their common velocity have been if they had been travelling in opposite directions before the collision? [2]

21 A tennis player strikes a ball so that its path is exactly reversed. The ball approaches the racket at $35\,\text{m s}^{-1}$ and leaves at $45\,\text{m s}^{-1}$. The mass of the ball is 90 grams. Find the magnitude of the impulse exerted on the ball. [2]

22 A gun of mass $500\,\text{kg}$, which is free to move, fires a shell of mass $5\,\text{kg}$ horizontally at a speed of $200\,\text{m s}^{-1}$.

   a Find the speed of recoil of the gun. [3]

   b In what ways would the situation change if the shell were fired at an elevation above the horizontal? [2]

23 Two particles A and B, of mass $4\,\text{kg}$ and $2\,\text{kg}$ respectively, are moving with respective speeds of $1\,\text{m s}^{-1}$ and $10\,\text{m s}^{-1}$ directly towards a fixed vertical wall. B hits the wall, rebounds and then collides with A. The coefficient of restitution between B and the wall is 0.4, and between the particles is 0.2

   a Show that B is brought to rest when it collides with A [3]

   b Calculate the speed at which A is travelling after its collision with B [2]

24 An object of mass $4\,\text{kg}$, travelling with velocity $(5\mathbf{i} + 2\mathbf{j})\,\text{m s}^{-1}$, is struck by a second object of mass $6\,\text{kg}$ and velocity $\mathbf{v}$. The two objects coalesce. Their common velocity after impact is $(2\mathbf{i} - 4\mathbf{j})\,\text{m s}^{-1}$. Find $\mathbf{v}$ [3]

25 A particle A, of mass $3\,\text{kg}$ and travelling at $10\,\text{m s}^{-1}$, catches up and collides with a second particle, B, of mass $5\,\text{kg}$, travelling at $2\,\text{m s}^{-1}$. After the collision, A is moving in the same direction with its speed reduced to $4\,\text{m s}^{-1}$. Calculate

   a The new speed of B, [2]

   b The coefficient of restitution. [2]

26 A particle of mass $6\,\text{kg}$, travelling at $8\,\text{m s}^{-1}$, is brought to rest in a head-on collision with a second particle of mass $4\,\text{kg}$. If $e = 0.3$, calculate the initial and final velocities of the second particle. [4]

27

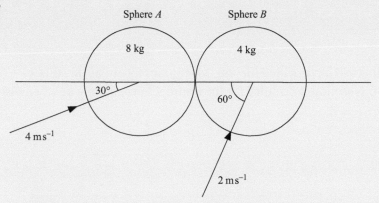

The diagram shows the oblique impact between spheres A and B, whose masses are 8 kg and 4 kg respectively. A has initial speed 4 m s^{-1} and B has initial speed 2 m s^{-1} at the angles shown. The coefficient of restitution between the spheres is 0.5. Find the speed and direction of travel of each sphere after the impact. [10]

28 A body of mass 5 kg is travelling with velocity $(2\mathbf{i} + 3\mathbf{j})$ m s^{-1} when it receives an impulse $\mathbf{I}$ N s. This changes its velocity to $(4\mathbf{i} + 7\mathbf{j})$ m s^{-1}. Find $\mathbf{I}$ and show that it has a magnitude of $10\sqrt{5}$ N s. [4]

29 A particle of mass 8 kg is travelling with velocity 2 m s^{-1} when it is acted upon for a period of 4 s by a forward force $F = (3t^2 + 1)$ N. Find

a The impulse received by the particle, [2]

b Its final velocity. [2]

30 A snooker ball, of mass 0.2 kg, hits the cushion with speed 10 m s^{-1}. Its direction makes an angle of 60° with the cushion. The coefficient of restitution is 0.8. How much kinetic energy is lost during the collision? [5]

31

Two identical spheres A and B of mass 2 kg collide. Sphere A has speed 6 m s^{-1} and B has speed 8 m s^{-1}. Their directions of travel make angles of 30° and 45° with the line of centres at the moment of impact, as shown. The coefficient of restitution between the spheres is 0.4. Find the loss of kinetic energy during the impact. [10]

Mathematics formulae
For AS and A Level Further Maths

The following mathematical formulae will be provided for you. Mechanics formulae, for this course, are provided at the start of this section. Formulae for the entire A Level course are then also provided for your reference.

Mechanics

Centres of mass

For uniform bodies:

Triangular lamina: $\dfrac{2}{3}$ along median from vertex

Circular arc, radius r, angle at centre 2α: $\dfrac{r\sin\alpha}{\alpha}$ from centre

Sector of circle, radius r, angle at centre 2α: $\dfrac{2r\sin\alpha}{3\alpha}$ from centre

Solid hemisphere, radius r: $\dfrac{3}{8}r$ from centre

Hemispherical shell, radius r: $\dfrac{1}{2}r$ from centre

Solid cone or pyramid of height h: $\dfrac{1}{4}h$ above the base on the line from centre of base to vertex

Conical shell of height h: $\dfrac{1}{3}h$ above the base on the line from centre of base to vertex

Motion in a circle

Transverse velocity: $v = r\dot{\theta}$

Transverse acceleration: $\dot{v} = r\ddot{\theta}$

Radial acceleration: $-r\dot{\theta}^2 = -\dfrac{v^2}{r}$

Full A Level

Pure Mathematics

Summations

$$\sum_{r=1}^{n} r^2 = \frac{1}{6}n(n+1)(2n+1) \qquad \sum_{r=1}^{n} r^3 = \frac{1}{4}n^2(n+1)^2$$

Matrix transformations

Anticlockwise rotation through θ about O: $\begin{pmatrix} \cos\theta & -\sin\theta \\ \sin\theta & \cos\theta \end{pmatrix}$

Reflection in the line $y = (\tan\theta)x$: $\begin{pmatrix} \cos2\theta & \sin2\theta \\ \sin2\theta & -\cos2\theta \end{pmatrix}$

Area of a sector

$A = \dfrac{1}{2}\int r^2\,d\theta$ \qquad (polar coordinates)

Complex numbers

$\{r(\cos\theta+i\sin\theta)\}^n = r^n(\cos n\theta+i\sin n\theta)$

The roots of $z^n = 1$ are given by $z = e^{\frac{2\pi ki}{n}}$ for $k = 0, 1, 2, ..., n-1$

Maclaurin's and Taylor's Series

$$f(x) = f(0) + xf'(0) + \frac{x^2}{2!}f''(0) + ... + \frac{x^r}{r!}f^{(r)}(0) + ...$$

$$e^x = \exp(x) = 1 + x + \frac{x^2}{2!} + ... + \frac{x^r}{r!} + ... \qquad \text{for all } x$$

$$\ln(1+x) = x - \frac{x^2}{2} + \frac{x^3}{3} - ... + (-1)^{r+1}\frac{x^r}{r} + ... \qquad (-1 < x \leq 1)$$

$$\sin x = x - \frac{x^3}{3!} + \frac{x^5}{5!} - ... + (-1)^r\frac{x^{2r+1}}{(2r+1)!} + ... \qquad \text{for all } x$$

$$\cos x = 1 - \frac{x^2}{2!} + \frac{x^4}{4!} - ... + (-1)^r\frac{x^{2r}}{(2r)!} + ... \qquad \text{for all } x$$

$$\arctan x = x - \frac{x^3}{3} + \frac{x^5}{5} - ... + (-1)^r\frac{x^{2r+1}}{2r+1} + ... \qquad (-1 \leq x \leq 1)$$

Vectors

Vector products: $\mathbf{a} \times \mathbf{b} = |\mathbf{a}||\mathbf{b}|\sin\theta\,\hat{\mathbf{n}} = \begin{vmatrix} \mathbf{i} & \mathbf{j} & \mathbf{k} \\ a_1 & a_2 & a_3 \\ b_1 & b_2 & b_3 \end{vmatrix} = \begin{pmatrix} a_2b_3 - a_3b_2 \\ a_3b_1 - a_1b_3 \\ a_1b_2 - a_2b_1 \end{pmatrix}$

$$\mathbf{a}\cdot(\mathbf{b}\times\mathbf{c}) = \begin{vmatrix} a_1 & a_2 & a_3 \\ b_1 & b_2 & b_3 \\ c_1 & c_2 & c_3 \end{vmatrix} = \mathbf{b}\cdot(\mathbf{c}\times\mathbf{a}) = \mathbf{c}\cdot(\mathbf{a}\times\mathbf{b})$$

If A is the point with position vector $\mathbf{a} = a_1\mathbf{i} + a_2\mathbf{j} + a_3\mathbf{k}$ and the direction vector $\mathbf{b}$ is given by $\mathbf{b} = b_1\mathbf{i} + b_2\mathbf{j} + b_3\mathbf{k}$, then the straight line through A with direction vector $\mathbf{b}$ has cartesian equation

$$\frac{x-a_1}{b_1} = \frac{y-a_2}{b_2} = \frac{z-a_3}{b_3}(=\lambda)$$

The plane through A with normal vector $\mathbf{n} = n_1\mathbf{i} + n_2\mathbf{j} + n_3\mathbf{k}$ has cartesian equation $n_1x + n_2y + n_3z + d = 0$ where $d = -\mathbf{a}.\mathbf{n}$

The plane through non-collinear points A, B and C has vector equation

$$\mathbf{r} = \mathbf{a} + \lambda(\mathbf{b} - \mathbf{a}) + \mu(\mathbf{c} - \mathbf{a}) = (1 - \lambda - \mu)\mathbf{a} + \lambda\mathbf{b} + \mu\mathbf{c}$$

The plane through the point with position vector $\mathbf{a}$ and parallel to $\mathbf{b}$ and $\mathbf{c}$ has equation

$$\mathbf{r} = \mathbf{a} + s\mathbf{b} + t\mathbf{c}$$

The perpendicular distance of (α, β, γ) from $n_1x + n_2y + n_3z + d = 0$ is $\dfrac{|n_1\alpha + n_2\beta + n_3\gamma + d|}{\sqrt{n_1^2 + n_2^2 + n_3^2}}$

Hyperbolic functions

$\cosh^2 x - \sinh^2 x = 1$

$\sinh 2x = 2 \sinh x \cosh x$

$\cosh 2x = \cosh^2 x + \sinh^2 x$

$\operatorname{arcosh} x = \ln\left\{x + \sqrt{x^2 - 1}\right\} \qquad (x \geq 1)$

$\operatorname{arsinh} x = \ln\left\{x + \sqrt{x^2 + 1}\right\}$

$\operatorname{artanh} x = \dfrac{1}{2}\ln\left(\dfrac{1+x}{1-x}\right) \qquad (|x| < 1)$

Conics

	Ellipse	Parabola	Hyperbola	Rectangular Hyperbola
Standard Form	$\dfrac{x^2}{a^2} + \dfrac{y^2}{b^2} = 1$	$y^2 = 4ax$	$\dfrac{x^2}{a^2} - \dfrac{y^2}{b^2} = 1$	$xy = c^2$
Parametric Form	$(a\cos\theta,\, b\sin\theta)$	$(at^2,\, 2at)$	$(a\sec\theta,\, b\tan\theta)$ $(\pm a\cosh\theta,\, b\sinh\theta)$	$\left(ct,\, \dfrac{c}{t}\right)$
Eccentricity	$e < 1$ $b^2 = a^2(1 - e^2)$	$e = 1$	$e > 1$ $b^2 = a^2(e^2 - 1)$	$e = \sqrt{2}$
Foci	$(\pm ae,\, 0)$	$(a,\, 0)$	$(\pm ae,\, 0)$	$\left(\pm\sqrt{2}c,\, \pm\sqrt{2}c\right)$
Directrices	$x = \pm\dfrac{a}{e}$	$x = -a$	$x = \pm\dfrac{a}{e}$	$x + y = \pm\sqrt{2}c$
Asymptotes	none	none	$\dfrac{x}{a} = \pm\dfrac{y}{b}$	$x = 0,\, y = 0$

Differentiation

$f(x)$	$f'(x)$		$f(x)$	$f'(x)$
$\arcsin x$	$\dfrac{1}{\sqrt{1-x^2}}$		$\cosh x$	$\sinh x$
$\arccos x$	$-\dfrac{1}{\sqrt{1-x^2}}$		$\tanh x$	$\operatorname{sech}^2 x$
$\arctan x$	$\dfrac{1}{1+x^2}$		$\operatorname{arsinh} x$	$\dfrac{1}{\sqrt{1+x^2}}$
$\sinh x$	$\cosh x$		$\operatorname{arcosh} x$	$\dfrac{1}{\sqrt{x^2-1}}$
			$\operatorname{artanh} x$	$\dfrac{1}{1-x^2}$

Integration (+ constant; $a > 0$ where relevant)

$f(x)$	$\int f(x)\,dx$
$\sinh x$	$\cosh x$
$\cosh x$	$\sinh x$
$\tanh x$	$\ln \cosh x$
$\dfrac{1}{\sqrt{a^2 - x^2}}$	$\arcsin\left(\dfrac{x}{a}\right) \quad (|x| < a)$
$\dfrac{1}{a^2 + x^2}$	$\dfrac{1}{a}\arctan\left(\dfrac{x}{a}\right)$
$\dfrac{1}{\sqrt{x^2 - a^2}}$	$\operatorname{arcosh}\left(\dfrac{x}{a}\right),\ \ln\{x + \sqrt{x^2 - a^2}\} \quad (x > a)$
$\dfrac{1}{\sqrt{a^2 + x^2}}$	$\operatorname{arsinh}\left(\dfrac{x}{a}\right),\ \ln\{x + \sqrt{x^2 + a^2}\}$
$\dfrac{1}{a^2 - x^2}$	$\dfrac{1}{2a}\ln\left|\dfrac{a + x}{a - x}\right| = \dfrac{1}{a}\operatorname{artanh}\left(\dfrac{x}{a}\right) \quad (|x| < a)$
$\dfrac{1}{x^2 - a^2}$	$\dfrac{1}{2a}\ln\left|\dfrac{x - a}{x + a}\right|$

Arc length

$$s = \int \sqrt{1 + \left(\frac{dy}{dx}\right)^2}\,dx \qquad \text{(cartesian coordinates)}$$

$$s = \int \sqrt{\left(\frac{dx}{dt}\right)^2 + \left(\frac{dy}{dt}\right)^2}\,dt \qquad \text{(parametric form)}$$

$$s = \int \sqrt{r^2 + \left(\frac{dr}{d\theta}\right)^2}\,d\theta \qquad \text{(polar form)}$$

Full A Level

Surface area of revolution

$$s_x = 2\pi \int y \sqrt{\left(1+\left(\frac{dy}{dx}\right)^2\right)}\, dx \qquad \text{(cartesian coordinates)}$$

$$s_x = 2\pi \int y \sqrt{\left(\frac{dx}{dt}\right)^2+\left(\frac{dy}{dt}\right)^2}\, dt \qquad \text{(parametric form)}$$

$$s_x = 2\pi \int r\sin\theta \sqrt{r^2+\left(\frac{dr}{d\theta}\right)^2}\, d\theta \qquad \text{(polar form)}$$

Statistics

Discrete distributions

For a discrete random variable X taking values x_i with probabilities $P(X = x_i)$

Expectation (mean): $E(X) = \mu = \Sigma x_i P(X = x_i)$

Variance: $\text{Var}(X) = \sigma^2 = \Sigma(x_i - \mu)^2\, P(X = x_i) = \Sigma\, x_i^2\, P(X = x_i) - \mu^2$

For a function $g(X)$: $E(g(X)) = \Sigma g(x_i)\, P(X = x_i)$

The probability generating function of X is $G_X(t) = E(t^X)$ and

$E(X) = G'_X(1)$ and $\text{Var}(X) = G''_X(1) + G'_X(1) - [G'_X(1)]^2$

For $Z = X + Y$, where X and Y are independent: $G_Z(t) = G_X(t) \times G_Y(t)$

Discrete distributions

Standard discrete distributions:

Distribution of X	$P(X = x)$	Mean	Variance	P.G.F.
Binomial $B(n, p)$	$\binom{n}{x}p^x(1-p)^{n-x}$	np	$np(1-p)$	$(1-p+pt)^n$
Poisson $\text{Po}(\lambda)$	$e^{-\lambda}\dfrac{\lambda^x}{x!}$	λ	λ	$e^{\lambda(t-1)}$
Geometric $\text{Geo}(p)$ on 1, 2, ...	$P(1-p)^{x-1}$	$\dfrac{1}{p}$	$\dfrac{1-p}{p^2}$	$\dfrac{pt}{1-(1-p)t}$
Negative binomial on $r, r+1, ...$	$\binom{x-1}{r-1}p^r(1-p)^{x-r}$	$\dfrac{r}{p}$	$\dfrac{r(1-p)}{p^2}$	$\left(\dfrac{pt}{1-(1-p)t}\right)^r$

Continuous distributions

For a continuous random variable X having probability density function f

Expectation (mean): $E(X) = \mu = \int x\,f(x)\,dx$

Variance: $\mathrm{Var}(X) = \sigma^2 = \int (x-\mu)^2\,f(x)\,dx = \int x^2\,f(x)\,dx - \mu^2$

For a function $g(X)$: $E(g(X)) = \int g(x)\,f(x)\,dx$

Cumulative distribution function: $F(x_0) = P(X \le x_0) = \int_{-\infty}^{x_0} f(t)\,dt$

Standard continuous distribution:

Distribution of X	P.D.F.	Mean	Variance
Normal $N(\mu, \sigma^2)$	$\dfrac{1}{\sigma\sqrt{2\pi}} e^{-\frac{1}{2}\left(\frac{x-\mu}{\sigma}\right)^2}$	μ	σ^2
Uniform (Rectangular) on $[a, b]$	$\dfrac{1}{b-a}$	$\dfrac{1}{2}(a+b)$	$\dfrac{1}{12}(b-a)^2$

Correlation and regression

For a set of n pairs of values (x_i, y_i)

$$S_{xx} = \sum(x_i - \bar{x})^2 = \sum x_i^2 - \frac{\left(\sum x_i\right)^2}{n}$$

$$S_{yy} = \sum(y_i - \bar{y})^2 = \sum y_i^2 - \frac{\left(\sum y_i\right)^2}{n}$$

$$S_{xy} = \sum(x_i - \bar{x})(y_i - \bar{y}) = \sum x_i y_i - \frac{\left(\sum x_i\right)\left(\sum y_i\right)}{n}$$

The product moment correlation coefficient is

$$r = \frac{S_{xy}}{\sqrt{S_{xx}S_{yy}}} = \frac{\sum(x_i - \bar{x})(y_i - \bar{y})}{\sqrt{\left\{\sum(x_i-\bar{x})^2\right\}\left\{\sum(y_i-\bar{y})^2\right\}}} = \frac{\sum x_i y_i - \frac{\left(\sum x_i\right)\left(\sum y_i\right)}{n}}{\sqrt{\left(\sum x_i^2 - \frac{\left(\sum x_i\right)^2}{n}\right)\left(\left(\sum y_i^2\right) - \frac{\left(\sum y_2\right)^2}{n}\right)}}$$

The regression coefficient of y on x is $b = \dfrac{S_{xy}}{S_{xx}} = \dfrac{\sum(x_i - \bar{x})(y_i - \bar{y})}{\sum(x_i - \bar{x})^2}$

Least squares regression line of y on x is $y = a + bx$ where $a = \bar{y} - b\bar{x}$

Residual Sum of Squares (RSS) $= S_{yy} - \dfrac{(S_{xy})^2}{S_{xx}} = S_{yy}(1 - r^2)$

Spearman's rank correlation coefficient is $r_s = 1 - \dfrac{6\sum d^2}{n(n^2 - 1)}$

Expectation algebra

For independent random variables X and Y

$$E(XY) = E(X)E(Y), \qquad \text{Var}(aX \pm bY) = a^2\,\text{Var}(X) + b^2\,\text{Var}(Y)$$

Sampling distributions

(i) Tests for mean when ρ is known

For a random sample $X_1, X_2, ..., X_n$ of n independent observations from a distribution having mean μ and variance σ^2:

$\overline{X}$ is an unbiased estimator of μ, with $\text{var}(\overline{X}) = \dfrac{\sigma^2}{n}$

S^2 is an unbiased estimator of σ^2, where $S^2 = \dfrac{\sum(X_i - \overline{X})^2}{n-1}$

For a random sample of n observation from $N(\mu, \sigma^2)$, $\dfrac{\overline{X} - \mu}{\sigma/\sqrt{n}} \sim N(0, 1)$

For a random sample of n_x observations from $N(\mu_x, \sigma_x^2)$, and, independently, a random sample of n_y observations from $N(\mu_y, \sigma_y^2)$, $\dfrac{(\overline{X} - \overline{Y}) - (\mu_x - \mu_y)}{\sqrt{\dfrac{\sigma_x^2}{n_x} + \dfrac{\sigma_y^2}{n_y}}} \sim N(0, 1)$

(ii) Tests for variance and mean when ρ is not known

For a random sample of n observations from $N(\mu, \sigma^2)$

$$\frac{(n-1)S^2}{\sigma^2} \sim X_{n-1}^2$$

$$\frac{\overline{X} - \mu}{S/\sqrt{n}} \sim t_{n-1} \text{ (Also valid in matched-pairs situations)}$$

For a random sample of n_x observations from $N(\mu_x, \sigma_x^2)$ and, independently, a random sample of n_y observations from $N(\mu_y, \sigma_y^2)$

$$\frac{S_x^2/\sigma_x^2}{S_y^2/\sigma_y^2} \sim F_{n_x-1, n_y-1,}$$

If $\sigma_x^2 = \sigma_y^2 = \sigma^2$ (unknown) then

$$\frac{\left(\overline{X}-\overline{Y}\right)-\left(\mu_x-\mu_y\right)}{\sqrt{S_P^2\left(\dfrac{1}{n_x}+\dfrac{1}{n_y}\right)}} \sim t_{n_x+n_y-2} \quad \text{where} \quad S_P^2 = \frac{(n_x-1)S_x^2+(n_y-1)S_y^2}{n_x+n_y-2}$$

Non parametric tests

Goodness-of-fit test and contingency tables: $\displaystyle\sum \frac{\left(O_i-E_i\right)^2}{E_i} \sim X_v^2$

You are expected to know the following Mathematical formulae, and they will not be provided for you. Mechanics formulae, for this course, are provided at the start. Formulae for the entire A Level course are then also provided for your reference.

Mechanics

Forces and Equilibrium

Weight $= \text{mass} \times \mathbf{g}$

Friction: $\mathbf{F} \leq \mu R$

Newton's second law in the form: $\mathbf{F} = m\mathbf{a}$

Kinematics

For motion in a straight line with variable acceleration:

$$\mathbf{v} = \frac{d\mathbf{r}}{dt} \qquad \mathbf{a} = \frac{d\mathbf{v}}{dt} = \frac{d^2\mathbf{r}}{dt^2}$$

$$\mathbf{r} = \int \mathbf{v}\,dt \qquad \mathbf{v} = \int \mathbf{a}\,dt$$

Momentum $= m\mathbf{v}$

Impulse $= m\mathbf{v} - m\mathbf{u}$

Kinetic energy $= \dfrac{1}{2}m\mathbf{v}^2$

Potential energy $= mgh$

The tension in an elastic string $= \dfrac{\lambda x}{l}$

The energy stored in an elastic string $= \dfrac{\lambda x^2}{2l}$

For Simple harmonic motion:

$$\ddot{x} = -\omega^2 x$$

$$x = a \cos \omega t \quad \text{or} \quad x = a \sin \omega t$$

$$v^2 = \omega^2(a^2 - x^2)$$

$$T = \frac{2\pi}{\omega}$$

Full A Level

Pure Mathematics

Quadratic Equations

$ax^2 + bx + c = 0$ has roots $\dfrac{-b \pm \sqrt{b^2 - 4ac}}{2a}$

Laws of indices

$a^x a^y \equiv a^{x+y}$

$a^x \div a^y \equiv a^{x-y}$

$(a^x)^y \equiv a^{xy}$

Laws of logarithms

$x = a^n \Leftrightarrow n = \log_a x$ for $a > 0$ and $x > 0$

$\log_a x + \log_a y \equiv \log_a xy$

$\log_a x - \log_a y \equiv \log_a \left(\dfrac{x}{y} \right)$

$k \log_a x \equiv \log_a (x)^k$

Coordinate geometry

A straight-line graph, gradient m passing through (x_1, y_1), has equation $y - y_1 = m(x - x_1)$

Straight lines with gradients m_1 and m_2 are perpendicular when $m_1 m_2 = -1$

Sequences

General term of an arithmetic progression:

$u_n = a + (n-1)d$

General term of a geometric progression:

$u_n = ar^{n-1}$

Trigonometry

In the triangle ABC:

Sine rule: $\dfrac{a}{\sin A} = \dfrac{b}{\sin B} = \dfrac{c}{\sin C}$

Cosine rule: $a^2 = b^2 + c^2 - 2bc \cos A$

Area $= \dfrac{1}{2} ab \sin C$

$\cos^2 A + \sin^2 A \equiv 1$

$\sec^2 A \equiv 1 + \tan^2 A$

$\operatorname{cosec}^2 A \equiv 1 + \cot^2 A$

$\sin 2A \equiv 2 \sin A \cos A$

$\cos 2A \equiv \cos^2 A - \sin^2 A$

$\tan 2A \equiv \dfrac{2 \tan A}{1 - \tan^2 A}$

Mensuration

Circumference and area of circle radius r and diameter d:

$C = 2\pi r = \pi d$ $\qquad$ $A = \pi r^2$

Pythagoras' Theorem:

In any right-angled triangle where a, b and c are the lengths of the sides and c is the hypotenuse, $c^2 = a^2 + b^2$

Area of a trapezium $= \dfrac{1}{2}(a+b)h$, where a and b are the lengths of the parallel sides and h is their perpendicular separation.

Volume of a prism $=$ area of cross section $\times$ length

For a circle of radius r, where an angle at the centre of θ radians subtends an arc of length s and encloses an associated sector of area A:

$s = r\theta$ $\qquad$ $A = \dfrac{1}{2} r^2 \theta$

Complex Numbers

For two complex numbers $z_1 = r_1 e^{i\theta_1}$ and $z_2 = r_2 e^{i\theta_2}$

$$z_1 z_2 = r_1 r_2\, e^{i(\theta_1 + \theta_2)}$$

$$\frac{z_1}{z_2} = \frac{r_1}{r_2} e^{i(\theta_1 - \theta_2)}$$

Loci in the Argand diagram:

$|z - a| = r$ is a circle radius r centred at a

$\arg(z - a) = \theta$ is a half line drawn from a at angle θ to a line parallel to the positive real axis.

Exponential Form: $e^{i\theta} = \cos\theta + i\sin\theta$

Matrices

For a 2 by 2 matrix $\begin{pmatrix} a & b \\ c & d \end{pmatrix}$ the determinant $\Delta = \begin{vmatrix} a & b \\ c & d \end{vmatrix} = ad - bc$

the inverse is $\dfrac{1}{\Delta} \begin{pmatrix} d & -b \\ -c & a \end{pmatrix}$

The transformation represented by matrix **AB** is the transformation represented by matrix **B** followed by the transformation represented by matrix **A**.

For matrices **A**, **B**:

$$(\mathbf{AB})^{-1} = \mathbf{B}^{-1}\mathbf{A}^{-1}$$

Algebra

$$\sum_{r=1}^{n} r = \frac{1}{2}n(n+1)$$

For $ax^2 + bx + c = 0$ with roots α and β:

$$\alpha + \beta = -\frac{b}{a} \qquad \alpha\beta = \frac{c}{a}$$

For $ax^3 + bx^2 + cx + d = 0$ with roots α, β and γ:

$$\sum \alpha = -\frac{b}{a} \qquad \sum \alpha\beta = \frac{c}{a} \qquad \alpha\beta\gamma = -\frac{d}{a}$$

Hyperbolic Functions

$$\cosh x \equiv \frac{1}{2}(e^x + e^{-x})$$

$$\sinh x \equiv \frac{1}{2}(e^x - e^{-x})$$

$$\tanh x \equiv \frac{\sinh x}{\cosh x}$$

Calculus and Differential Equations

<u>Differentiation</u>

Function	Derivative
x^n	nx^{n-1}
$\sin kx$	$k \cos kx$
$\cos kx$	$-k \sin kx$
$\sinh kx$	$k \cosh kx$
$\cosh kx$	$k \sinh kx$
e^{kx}	ke^{kx}
$\ln x$	$\dfrac{1}{x}$
$f(x) + g(x)$	$f'(x) + g'(x)$
$f(x)g(x)$	$f'(x)g(x) + f(x)g'(x)$
$f(g(x))$	$f'(g(x))g'(x)$

Integration

Function	Integral
x^n	$\dfrac{1}{n+1}x^{n+1}+c,\ n\neq-1$
$\cos kx$	$\dfrac{1}{k}\sin kx+c$
$\sin kx$	$-\dfrac{1}{k}\cos kx+c$
$\cosh kx$	$\dfrac{1}{k}\sinh kx+c$
$\sinh kx$	$\dfrac{1}{k}\cosh kx+c$
e^{kx}	$\dfrac{1}{k}e^{kx}+c$
$\dfrac{1}{x}$	$\ln|x|+c,\ x\neq0$
$f'(x)+g'(x)$	$f(x)+g(x)+c$
$f'(g(x))g'(x)$	$f(g(x))+c$

$$\text{Area under a curve} = \int_a^b y\,dx\ (y\geq0)$$

Volumes of revolution about the x and y axes:

$$V_x=\pi\int_a^b y^2\,dx \qquad V_y=\pi\int_c^d x^2\,dy$$

Simple Harmonic Motion:

$$\ddot{x}=-\omega^2 x$$

Vectors

$$|x\mathbf{i} + y\mathbf{j} + z\mathbf{k}| = \sqrt{(x^2 + y^2 + z^2)}$$

Scalar product of two vectors $\mathbf{a} = \begin{pmatrix} a_1 \\ a_2 \\ a_3 \end{pmatrix}$ and $\mathbf{b} = \begin{pmatrix} b_1 \\ b_2 \\ b_3 \end{pmatrix}$ is

$$\begin{pmatrix} a_1 \\ a_2 \\ a_3 \end{pmatrix} \cdot \begin{pmatrix} b_1 \\ b_2 \\ b_3 \end{pmatrix} = a_1 b_1 + a_2 b_2 + a_3 b_3 = |\mathbf{a}| \, |\mathbf{b}| \cos \theta$$

where θ is the acute angle between the vectors $\mathbf{a}$ and $\mathbf{b}$

The equation of the line through the point with position vector $\mathbf{a}$ parallel to vector $\mathbf{b}$ is:

$$\mathbf{r} = \mathbf{a} + t\mathbf{b}$$

The equation of the plane containing the point with position vector $\mathbf{a}$ and perpendicular to vector $\mathbf{n}$ is:

$$(\mathbf{r} - \mathbf{a}) \cdot \mathbf{n} = 0$$

Statistics

The mean of a set of data: $\overline{x} = \dfrac{\sum x}{n} = \dfrac{\sum fx}{\sum f}$

The standard Normal variable: $Z = \dfrac{X - \mu}{\sigma}$ where $X \sim N(\mu, \sigma^2)$

Mathematical notation
For AS and A Level Further Maths

You should understand the following notation for AS and A Level Further Maths, without need for further explanation. Notation from all strands of mathematics has been included for your reference.

Mechanics

kg	kilograms
m	metres
km	kilometres
m/s, m s^{-1}	metres per second (velocity)
m/s^2, m s^{-2}	metres per second squared (acceleration)
F	Force or resultant force
N	Newton
Nm	Newton metre (moment of a force)
t	time
s	displacement
u	initial velocity
v	velocity or final velocity
a	acceleration
g	acceleration due to gravity
μ	coefficient of friction

Set Notation

$\in$	is an element of
$\notin$	is not an element of
$\subseteq$	is a subset of
$\subset$	is a proper subset of
$\{x_1, x_2,... \}$	the set with elements x_1, x_2, ...
$\{x: ... \}$	the set of all x such that ...
$n(A)$	the number of elements in set A
$\varnothing$	the empty set
ε	the universal set
A'	the complement of the set A
$\mathbb{N}$	the set of natural numbers, $\{1, 2, 3, ...\}$
$\mathbb{Z}$	the set of integers, $\{0, \pm 1, \pm 2, \pm 3, ...\}$
$\mathbb{Z}^+$	the set of positive integers, $\{1, 2, 3, ...\}$
$\mathbb{Z}_0^+$	the set of non-negative integers, $\{0, 1, 2, 3, ...\}$
$\mathbb{R}$	the set of real numbers
$\mathbb{Q}$	the set of rational numbers, $\left\{ \dfrac{p}{q} : p \in \mathbb{Z},\ q \in \mathbb{Z}^+ \right\}$
$\cup$	union
$\cap$	intersection
(x, y)	the ordered pair x, y
$[a, b]$	the closed interval $\{x \in \mathbb{R} : a \leq x \leq b\}$
$[a, b)$	the interval $\{x \in \mathbb{R} : a \leq x < b\}$
$(a, b]$	the interval $\{x \in \mathbb{R} : a < x \leq b\}$
(a, b)	the open interval $\{x \in \mathbb{R} : a < x < b\}$
$\mathbb{C}$	the set of complex numbers

Miscellaneous Symbols

$=$	is equal to
$\neq$	is not equal to
$\equiv$	is identical to or is congruent to
$\approx$	is approximately equal to
∞	infinity
$\propto$	is proportional to
$<$	is less than
$\leqslant, \leq$	is less than or equal to; is not greater than
$>$	is greater than
$\geqslant, \geq$	is greater than or equal to; is not less than
$\therefore$	therefore
$\because$	because
$p \Rightarrow q$	p implies q (if p then q)
$p \Leftarrow q$	p is implied by q (if q then p)
$p \Leftrightarrow q$	p implies and is implied by q (p is equivalent to q)
a	first term for an arithmetic or geometric sequence
l	last term for an arithmetic sequence
d	common difference for an arithmetic sequence
r	common ratio for a geometric sequence
S_n	sum to n terms of a sequence
S_∞	sum to infinity of a sequence

Operations

$a + b$	a plus b
$a - b$	a minus b
$a \times b, ab, a \cdot b$	a multiplied by b
$a \div b, \dfrac{a}{b}$	a divided by b
$\displaystyle\sum_{i=1}^{n} a_i$	$a_1 + a_2 + \ldots + a_n$
$\displaystyle\prod_{i=1}^{n} a_i$	$a_1 \times a_2 \times \ldots \times a_n$
$\sqrt{a}$	the non-negative square root of a
$|a|$	the modulus of a
$n!$	n factorial: $n! = n \times (n-1) \times \ldots \times 2 \times 1$, $n \in \mathbb{N}$; $0! = 1$
$\dbinom{n}{r}, {}^nC_r, {}_nC_r$	the binomial coefficient $\dfrac{n!}{r!(n-r)!}$ for $n, r \in \mathbb{Z}_0^+$, $r \leq n$ or $\dfrac{n(n-1)\ldots(n-r+1)}{r!}$ for $n \in \mathbb{Q}, r \in \mathbb{Z}_0^+$

Functions

$f(x)$	the value of the function f at x
$\displaystyle\lim_{x \to a} f(x)$	the limit of $f(x)$ as x tends to a
$f: x \mapsto y$	the function f maps the element x to the element y
f^{-1}	the inverse function of the function f
gf	the composite function of f and g which is defined by $gf(x) = g(f(x))$

Mathematical notation for AS and A Level Further Maths

$\Delta x, \delta x$ an increment of x

$\dfrac{dy}{dx}$ the derivative of y with respect to x

$\dfrac{d^n y}{dx^n}$ the nth derivative of y with respect to x

$f'(x) ..., f^{(n)}(x)$ the first, ..., nth derivatives of $f(x)$ with respect to x

$\dot{x}, \ddot{x}, ...$ the first, second, ... derivatives of x with respect to t

$\int y \, dx$ the indefinite integral of y with respect to x

$\int_a^b y \, dx$ the definite integral of y with respect to x between the limits $x = a$ and $x = b$

Exponential and Logarithmic Functions

e   base of natural logarithms

e^x, exp x exponential function of x

$\log_a x$ logarithm to the base a of x

$\ln x, \log_e x$ natural logarithm of x

Trigonometric Functions

$\left.\begin{array}{l} \text{sin, cos, tan,} \\ \text{cosec, sec, cot} \end{array}\right\}$ the trigonometric functions

$\left.\begin{array}{l} \sin^{-1}, \cos^{-1}, \tan^{-1} \\ \text{arcsin, arccos, arctan} \end{array}\right\}$ the inverse trigonometric functions

$\circ$ degrees

rad   radians

$\left.\begin{array}{l} \mathrm{cosec}^{-1}, \sec^{-1}, \cot^{-1} \\ \text{arccosec, arcsec, arccot} \end{array}\right\}$ the inverse trigonometric functions

$\left.\begin{array}{l} \text{sinh, cosh, tanh} \\ \text{cosech, sech, coth} \end{array}\right\}$ the hyperbolic functions

$\left.\begin{array}{l} \sinh^{-1}, \cosh^{-1}, \tanh^{-1} \\ \mathrm{cosech}^{-1}, \mathrm{sech}^{-1}, \coth^{-1} \\ \text{arsinh, arcosh, artanh} \\ \text{arcosech, arsech, arcoth} \end{array}\right\}$ the inverse hyperbolic functions

Complex numbers

i, j square root of -1

$x + iy$ complex number with real part x and imaginary part y

$r(\cos \theta + i \sin \theta)$ modulus argument form of a complex number with modulus r and argument θ

z a complex number $z = x + iy = r(\cos \theta + i \sin \theta)$

$\mathrm{Re}(z)$ the real part of z, $\mathrm{Re}(z) = x$

$\mathrm{Im}(z)$ the imaginary part of z, $\mathrm{Im}(z) = y$

$|z|$ the modulus of z, $|z| = \sqrt{x^2 + y^2}$

$\arg(z)$ the argument of z, $\arg(z) = \theta$, $-\pi < \theta < \pi$

z^* the complex conjugate of z, $x - iy$

Matrices

$\mathbf{M}$ a matrix $\mathbf{M}$

$\mathbf{0}$ zero matrix

$\mathbf{I}$ identity matrix

$\mathbf{M}^{-1}$ the inverse of the matrix $\mathbf{M}$

$\mathbf{M}^{\mathrm{T}}$ the transpose of the matrix $\mathbf{M}$

$\det \mathbf{M}$ or $|\mathbf{M}|$ the determinant of the square matrix $\mathbf{M}$

$\mathbf{Mr}$ Image of the column vector $\mathbf{r}$ under the transformation associated with the matrix $\mathbf{M}$

Vectors

$\mathbf{a}, \underline{a}, \underset{\sim}{a}$ the vector $\mathbf{a}, \underline{a}, \underset{\sim}{a}$

$\overrightarrow{AB}$ the vector represented in magnitude and direction by the directed line segment AB

$\hat{\mathbf{a}}$ a unit vector in the direction of $\mathbf{a}$

$\mathbf{i}, \mathbf{j}, \mathbf{k}$ unit vectors in the directions of the Cartesian coordinate axes

$|\mathbf{a}|, a$ the magnitude of $\mathbf{a}$

$|\overrightarrow{AB}|, AB$ the magnitude of $\overrightarrow{AB}$

$\begin{pmatrix} a \\ b \end{pmatrix}, a\mathbf{i} + b\mathbf{j}$ column vector and corresponding unit vector notation

$\mathbf{r}$ position vector

$\mathbf{s}$ displacement vector

$\mathbf{v}$ velocity vector

$\mathbf{a}$ acceleration vector

$\mathbf{a} \cdot \mathbf{b}$ the scalar product of $\mathbf{a}$ and $\mathbf{b}$

$\mathbf{a} \times \mathbf{b}$ the vector product of $\mathbf{a}$ and $\mathbf{b}$

$\mathbf{a}.\mathbf{b} \times \mathbf{c}$ the scalar triple product of $\mathbf{a}$, $\mathbf{b}$ and $\mathbf{c}$

Differential equations

ω angular speed

Probability and Statistics

A, B, C, etc.   events

$A \cup B$ union of the events A and B

$A \cap B$ intersection of the events A and B

$\mathrm{P}(A)$ probability of the event A

A' complement of the event A

$\mathrm{P}(A \mid B)$ probability of the event A conditional on the event B

X, Y, R, etc.   random variables

x, y, r, etc.   values of the random variables X, Y, R etc.

$x_1, x_2, \ldots$	observations
$f_1, f_2, \ldots$	frequencies with which the observations $x_1, x_2, \ldots$ occur
$p(x)$, $P(X = x)$	probability function of the discrete random variable X
$p_1, p_2, \ldots$	probabilities of the values $x_1, x_2, \ldots$ of the discrete random variable X
$E(X)$	expectation of the random variable X
$Var(X)$	variance of the random variable X
$\sim$	has the distribution
$B(n, p)$	binomial distribution with parameters n and p, where n is the number of trials and p is the probability of success in a trial
q	$q = 1 - p$ for binomial distribution
$N(\mu, \sigma^2)$	Normal distribution with mean μ and variance σ^2
$Z \sim N(0,1)$	standard Normal distribution
ϕ	probability density function of the standardised Normal variable with distribution $N(0, 1)$
Φ	corresponding cumulative distribution function
μ	population mean
σ^2	population variance
σ	population standard deviation
$\bar{x}$	sample mean
s^2	sample variance
s	sample standard deviation
H_0	Null hypothesis
H_1	Alternative hypothesis
r	product moment correlation coefficient for a sample
ρ	product moment correlation coefficient for a population
$Po(\lambda)$	Poisson distribution with parameter λ
$Geo(P)$	geometric distribution with parameter p
$G_X(t)$	probability generating function of the random variable X
χ^2_ν	chi squared distribution with ν degrees of freedom
t_n	t distribution with n degrees of freedom
F_{ν_1, ν_2}	F distribution with ν_1 and ν_2 degree of freedom

Answers

Exercise 1.1A

1 a   720 J, 180 W
   b   60 J, 40 W
2 a   16 J
   b $42\frac{2}{3}$ J
   c $21\frac{1}{3}$ J
3 a   30 J
   b   46 J
4 Box A
   a $12g$ N
   b $36g$ J
   c $9g$ W
   d $36g$ J

   Box B
   a $4g$ N
   b $2g$ J
   c $0.5g$ W
   d $2g$ J

   Box C
   a $0.5g$ N
   b $0.1g$ J
   c $0.05g$ W
   d $0.1g$ J
5 a   i 10^5 J   ii 10^5 J   iii   250 N
   b   120 000 J, 300 N
6 a 7.5×10^6 J
   b 8.5×10^6 J, 17 000 N
   c   170 000 W, 340 000 W
7 a   i 352.8 J, 352.8 J ii $24.2 \, \text{m s}^{-1}$
   b   470.4 J, $28 \, \text{m s}^{-1}$
   c $28 \, \text{m s}^{-1}$
   d $20.9 \, \text{m s}^{-1}$
8 a   i 1 764 000 J   ii   5880 N   iii   147 kW
   b   8880 N, 222 kW

Exercise 1.1B

1 a   375 kW   b $\frac{1}{4} \, \text{m s}^{-2}$
   c   In part **a**, the work done by the engine would now need to overcome the resistance as well as increase the KE. The energy equation would then be: Work done by engine = Increase in KE + Work done against the resistance. Part **b** would have the same answer as the velocities and distance travelled are unchanged.
2 435 000 W
3 a $3.55 \, \text{m s}^{-1}$
   b   Air resistance is not being considered; The pulley may not be smooth.
4 3640 J, 13.2 N
5 $26.7 \, \text{m s}^{-1}$
6 3560 W
7 1134 W
8 a $30 \, \text{m s}^{-1}$
   b $\frac{2}{3} \, \text{m s}^{-2}$
9 $21.5 \, \text{m s}^{-1}$
10 $\frac{81}{320}$
11 $0.147 \, \text{m s}^{-2}$
12 a $3.32 \, \text{m s}^{-1}$, 66.5 W
    b $s \leq 6$; A different model could be $T = 36 - 3s$ or $T = 100 - s^2$

13 a $8.24 \, \mathrm{m\,s^{-1}}$ **b** $7.38 \, \mathrm{m\,s^{-1}}$

   c **i** The value of v would be lower as energy would be expended in overcoming friction.

      ii No; resistance to motion would include air resistance, which would likely increase as the speed of the skateboard increases. A possible alternative would be $R = 48 + kv$, where k is a constant.

14 $20 \, \mathrm{m\,s^{-1}}$

15 a $0.391 \, \mathrm{m\,s^{-2}}$ **b** $207 \, \mathrm{N}$

16 a Energy equation from bottom to top of loop is

      GPE gained = KE lost

$$m \times g \times 2r = \frac{1}{2} m u^2$$

$$u^2 = 4gr$$

   b Energy equation from bottom to point P is

      GPE gained = KE lost

$$mgr + mgr \cos\theta = \frac{1}{2} m u^2 - \frac{1}{2} m \left(\frac{u}{2} \right)^2$$

$$gr(1 + \cos\theta) = \frac{3u^2}{8}$$

$$\cos\theta = \frac{3u^2}{8gr} - 1$$

      θ is acute, so $\cos\theta$

$$0 < \cos\theta < 1$$

$$0 < \frac{3u^2}{8gr} - 1 < 1$$

$$8gr \times 1 < 3u^2 \quad < 2 \times 8gr$$

$$\frac{8gr}{3} < u^2 \quad < \frac{16gr}{3}$$

17 $1.05°$

18 $65.3 \, \mathrm{m\,s^{-1}}$

19 $2.08 \, \mathrm{m\,s^{-1}}$, $41.5 \, \mathrm{W}$

20 a $1.5 \times 10^7 \, \mathrm{J}$

   b $8.40 \, \mathrm{m\,s^{-1}}$, $756 \, 312 \, \mathrm{W}$

   c The calculation of the work done uses an approximate method. The resistance is constant, though part of any resistance, such as air resistance, varies with speed.

   d The area under the s–T graph could be approximated by rectangles, trapezia or parabolas. The use of parabolas is likely to be the most accurate as it takes into account the curvature of the s–T graph.

Exercise 1.2A

1 **A** $2 \, \mathrm{N}$, $0.5 \, \mathrm{J}$ **B** $2 \, \mathrm{N}$, $1 \, \mathrm{J}$ **C** $0.4 \, \mathrm{m}$, $0.8 \, \mathrm{J}$ **D** $9 \, \mathrm{N}$, $1.35 \, \mathrm{J}$ **E** $0.8 \, \mathrm{m}$, $3.2 \, \mathrm{N}$

2 $0.4 \, \mathrm{m}$

3 $2.5 \, \mathrm{m}$

4 a $0.8 \, \mathrm{m}$ **b** $8 \, \mathrm{J}$ **c** $8 \, \mathrm{J}$

5 a $0.72 \, \mathrm{m}$ **b** $5.4 \, \mathrm{J}$

6 a $78.4 \, \mathrm{N}$ **b** $0.98 \, \mathrm{m}$

7 $x_1 = 1 \, \mathrm{m}$, $x_2 = 0.5 \, \mathrm{m}$, $26\frac{2}{3} \, \mathrm{N}$

8 $x_1 = \frac{4}{9} \, \mathrm{m}$, $x_2 = \frac{5}{9} \, \mathrm{m}$, $4.17 \, \mathrm{J}$

9 $0.182 \, \mathrm{m}$, $8.33 \, \mathrm{J}$

10 For vertical equilibrium

$$T_1 + T_2 = mg$$

$$\frac{\lambda_1}{l} x + \frac{\lambda_2}{l} x = mg$$

$$(\lambda_1 + \lambda_2)x = mgl$$

     Extension $x = \dfrac{mgl}{\lambda_1 + \lambda_2}$

     $\dfrac{mg\lambda_1}{\lambda_1 + \lambda_2} \, \mathrm{N}$, $\dfrac{mg\lambda_2}{\lambda_1 + \lambda_2} \, \mathrm{N}$

Exercise 1.2B

1 $16\frac{2}{3} \, \mathrm{J}$

2 a $39.2 \, \mathrm{N}$ in both   **b** $0.441 \, \mathrm{m}$ **c** $8.64 \, \mathrm{J}$

3 a $0.65 \, \mathrm{m}$ **b** $15 \, \mathrm{N}$, $34 \, \mathrm{N}$ **c** $15.6 \, \mathrm{J}$

4 a 980, 1.36 cm   **b** $3a$

5 6.62 m s^{-2}, 1.94 m

6 Work done = change in elastic potential energy

$$= \text{final EPE} - \text{initial EPE}$$

$$= \frac{1}{2}\frac{\lambda x_2^2}{l} - \frac{1}{2}\frac{\lambda x_1^2}{l} = \frac{\lambda}{2l}(x_2^2 - x_1^2)$$

$$= \frac{\lambda}{2l}(x_2 + x_1)(x_2 - x_1)$$

$$= \text{increase in extension} \times \frac{1}{2}\left(\frac{\lambda x_2}{l} + \frac{\lambda x_1}{l}\right)$$

$$= \text{increase in extension} \times \text{mean of initial and final tension}$$

$$= \frac{T_1 + T_2}{2}(x_2 - x_1)$$

Or using calculus

$$\text{Work done} = \int_{x_2}^{x_1} T \, dx$$

$$= \frac{\lambda}{l}\left[\frac{x^2}{2}\right]_{x_1}^{x_2}$$

$$= \frac{1}{2} \times \frac{\lambda}{l} \times \left(x_2^2 - x_1^2\right)$$

$$= \frac{1}{2} \times \frac{\lambda}{l}(x_2 + x_1)(x_2 - x_1)$$

$$= \frac{T_1 + T_2}{2}(x_2 - x_1)$$

Or using the graph of $T = \dfrac{\lambda x}{l}$

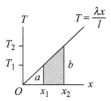

Work done = shaded area on T-x graph

$$= \text{area of a trapezium}$$

$$= \frac{1}{2}h(a + b)$$

$$= \frac{1}{2}(T_1 + T_2)(x_2 - x_1)$$

7 a 0.8 m

b 0.737 m

8 a 0.3 m

b 1.71 m s^{-1}, 0.3 m above P

The mass now falls back to Q and continues to oscillate about P, 0.3 m above and below.

c 1.23 m below A

d The mass will oscillate about P, 0.2 m above and below.

1.3 m below A

9 a Energy equation gives

$$\text{KE gained} + \text{EPE gained} = \text{GPE lost}$$

$$\frac{1}{2} \times 60 \times v^2 + \frac{1}{2} \times \frac{1200}{25}(x - 25)^2 = 60 \times g \times x$$

$$30v^2 + 24(x^2 - 50x + 625) = 588x$$

$$30v^2 = 1788x - 24x^2 - 15\,000$$

25, 24.7 m s^{-1}

b 64.9 m

c E.g. Include the weight of the rope; include the effect of air resistance.

10 a 2.235 m

b 0.509 m

c A smooth plane has no friction. So the equation along the plane for equilibrium is $T = 4g \sin 30°$, giving only one possible value of T and, from Hooke's Law, the extension x. The range of possible answers is reduced to this one value of x

11 a 3.13 m s^{-1}

   b Resolving vertically for equilibrium

$$2T \cos\theta = 15g$$

where $\theta = \tan^{-1}\left(\dfrac{2}{1.1}\right) = 61.1892\ldots°$

Hooke's law gives $T = \dfrac{30g}{1.5}\left(\sqrt{2^2 + 1.1^2} - 1.5\right)$

$$= 153.3783\ldots \text{ N}$$

$$2T \cos\theta = 2 \times 153.3783\ldots \times \cos 61.1892\ldots°$$

$$= 147.8317 \text{ N}$$

$$15g = 15 \times 9.8 = 147 \text{ N}$$

$147 \approx 147.8$, mass hangs in equilibrium

12 a 10.2 kg **b** $117°$ **c** Model the rod and string as having weight.

Review exercise 1

1 1080 N

2 a 150 N **b** $15\,000 \text{ J}$ **c** $15\,000 \text{ J}$ **d** 1500 W

3 a 0.24 m s^{-2} **b** 2.4 m s^{-1} **c** 576 J **d** $23\,520 \text{ J}$

   e $24\,096 \text{ J}$ **f** 2008 N **g** 4819.2 W

4 655 N

5 802.5 W

6 63 N

7 7.43 m s^{-1}, 156 W

8 If slope is at angle α to horizontal, and car travels distance s along slope at maximum speed, then, going down slope.

Work done against resistance = work done by engine + change in PE

$$kVs = \frac{Ps}{V} + mgs \sin\alpha$$

$$kV = \frac{P}{V} + mg \sin\alpha$$

Going up slope:

Work done by engine = work done against resistance + change in PE

$$\frac{2Ps}{V} = \frac{kVs}{2} + mgs \sin\alpha$$

$$\frac{2P}{V} = \frac{kV}{2} + mg \sin\alpha$$

So

$$kV - \frac{P}{V} = \frac{2P}{V} - \frac{kV}{2}$$

$$\frac{3kV}{2} = \frac{3P}{V}$$

$$V^2 = \frac{2P}{k}$$

$$V = \sqrt{\frac{2P}{k}}$$

OR

Newton's 2nd law gives:

Down slope

$$\frac{p}{v} + mg \sin\alpha - kv = 0$$

Up slope

$$\frac{2p}{v} - mg \sin\alpha - k\frac{v}{2} = 0$$

Add equations

$$\frac{3p}{v} - 3\frac{kv}{2} = 0$$

$$v^2 = 2\frac{p}{k}$$

$$v = \sqrt{\frac{2p}{k}}$$

9 2.40×10^{10} J

10 a 117.6 N

   b 1.47 m

   c 86.436 J

11 0.652 m, 0.348 m, 17.4 N

12 a 705 000 J

   b 1410 N

   c 35 250 W

13

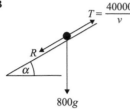

If it travels distance 1m at max velocity V, then

Work done by engine = work done against resistance + change in GPE

$$\frac{40000}{V} = (34 + 5V^2) + \frac{800g}{140}$$

OR

Newton's 2nd law gives:

$T - mg \sin \alpha - R = m \times 0$

As acceleration is zero at max speed

$$\frac{40000}{v} - 800g \frac{1}{140} - \left(34 + 5v^2\right) = 0$$

$$5v^3 + 90v - 40000 = 0$$

Substitute $v = 19.7$

$5 \times 19.7^3 + 90 \times 19.7 - 40000 = -0.135 \approx 0$

So max velocity = 19.7 m s^{-1}

2.10 m s^{-1}

14 a 2.48 m s^{-1}

   b Pythagoras gives

$$AM = \sqrt{9 + 1.76^2} = 3.4781... \text{ m}$$

and for equilibrium

$2T \cos \alpha = mg$

where $T = \dfrac{\lambda}{l} x$

$\qquad = \dfrac{20g}{2} \times 1.48$

$\qquad = 144.8597... \text{ N}$

$2T \cos \alpha = 2 \times 144.8597... \times \dfrac{1.76}{3.48}$

$\qquad = 146.6022 \approx 147 \text{ N}$

and $15g = 147$ N

Hence, mass is in equilibrium.

1 a 1176 J

   b 6.53 s

   c **i** The weight on the 'up side' will continually reduce, and on the 'down side' it will wither stay the same after some point (if the rope rests on the ground) or continually increase (if, for example, the rope continues to descend down the side of a scaffold). The final GPE of the rope will be less than its initial GPE, so the total work done will be less.

      ii (E.g.) The man must successively reposition his hands on the rope and this process will interrupt the smooth progress of the rope, so he will not be exerting constant power.

2 a 8.85 m s^{-1}

   b **i** 530 J   **ii** 88.3 N

3 a 33 N

   b 528 J

4 1.65 m, 0.745 m

5 4.70 m s^{-1}

6 3.83 m

7 14.5 kW

8 0.221 m s^{-2}

9 **a** 7.67 m s^{-1}

   b 4.17 m

   c **i** The ball is a particle/there is no air resistance.

   ii The speed reached and the total distance fallen would both be less.

10 a 1.1 m

   b 0.6 m

   c 0.2 m above A

11 a 3.97 m

   b 8.82 m s^{-1}

12 a 28.5 m s^{-1}

   b For example, firing upwards would achieve a slightly lower maximum speed than firing downwards, as the weight would hinder the acceleration in the first case and help it in the second. The effects would be insignificant because the GPE changes involved (around 0.02 J either way) are tiny in relation to the EPE (4.07 J) of the string.

13 a 0.395 m

   b 3.43 m s^{-1}

14 $8P$ W

15 400 N

16 108 km h^{-1}

17 5.21 m s^{-1}

18 0.170 m

19 a 12.2

   b 0.196 m s^{-2}

   c 31.4 m s^{-1}

20 16.8 m s^{-1}

21 a 866 J

   b 5.22 m s^{-1}

   c The string will have a finite breaking strain, so the force that can be applied is limited. A tension of more than $20\,g$ N would make the body leave the surface, so the problem would change.

Exercise 2.1A

1 a $2\frac{1}{2}$ m s^{-1}

   b 1 m s^{-1}

   c $4\frac{1}{2}$ m s^{-1}

   d −2 m s^{-1}

   e $-1\frac{1}{2}$ m s^{-1}

2 a 3.6 m s^{-1}

   b 1.5 m s^{-1}

   c −1.4 m s^{-1}

3 10 kg

4 1.57 m s^{-1}

Exercise 2.1B

1 20 m s^{-1}

2 0.676 m s^{-1}

3 6.43 km h^{-1}

4 11.8 m s^{-1}

5 a 3.2 m s^{-1}

   b 0.8 m s^{-1}

6 6.25 m s^{-1}

Taking gun and bullet together, the explosive impulse is not an external impulse, so does not affect the total momentum.

7 227 m s^{-1}

8 a $7\frac{1}{2}$ m s^{-1}

   b No account is taken of any resistance to motion between collisions, such as friction or air resistance.

9 a 5 m s^{-1}, $1\frac{2}{3}$ m s^{-1}

   There are no more collisions as X and Y together are moving in the opposite direction to Z

   b E.g. Motion takes place in a horizontal line unaffected by gravity; No account is taken of any resistance to motion between collisions; The particles are point masses or spherical masses of the same diameter without any spin.

10 a $3\frac{1}{3} \text{ms}^{-1}$

  b 2 ms^{-1}

   R is now moving towards S at rest so there is another collision between R and S

  c A more realistic model could take into account the weight of the string or its elasticity.

11 $\frac{2}{5}u$ and $\frac{4}{5}u$

12 Momentum equation gives

$$10 \times 5 + m \times (-2) = 10 \times 3 + mv$$
$$mv = 20 - 2m$$

No further collisions if
$$v \geq 3$$
$$\frac{20 - 2m}{m} \geq 3$$
$$20 \geq 5m$$
$$m \leq 4$$

Exercise 2.2A

1 a $3 \text{ ms}^{-1}, 4 \text{ ms}^{-1}$

  b $-3\frac{1}{3} \text{ ms}^{-1}, 2\frac{2}{3} \text{ ms}^{-1}$

  c $11 \text{ ms}^{-1}, 5 \text{ ms}^{-1}$

  d $-3 \text{ ms}^{-1}, 0 \text{ ms}^{-1}$

  e $-\frac{1}{3} \text{ ms}^{-1}, 2\frac{1}{6} \text{ ms}^{-1}$

2 a $11 \text{ ms}^{-1}, \frac{3}{4}$

  b $3.25 \text{ ms}^{-1}, \frac{1}{4}$

  c $0 \text{ ms}^{-1}, \frac{1}{5}$

  d $2 \text{ ms}^{-1}, \frac{4}{5}$

3 $0.18 \text{ ms}^{-1}, 0.15 \text{ ms}^{-1}$

4 $3 \text{ ms}^{-1}, 12 \text{ ms}^{-1}$

5 75.3 J

6 $e = \frac{2\sqrt{5}}{5} \ (= 0.894), 0.225 \text{ J}$

7 a $\frac{11}{15} \ (= 0.733)$

  b 0.000192 J

Exercise 2.2B

1 a $8.85 \text{ ms}^{-1}, 7.67 \text{ ms}^{-1}$

  b $\frac{\sqrt{3}}{2} = 0.866$

  c $9.8m \text{ J}$

  d $e = 0$ gives the speed of separation as zero, so the clay does not rebound. All the KE before impact is lost.

2 a $3\frac{1}{3} \text{ ms}^{-1}$

  b $83\frac{1}{3} \%$

3 a 2 kg

  b Y loses the least.

4 a **i** 1.5 kg **ii** 57.8%

  b There are no resistances to motion between collisions, e.g. the plane is smooth, air-resistance is negligible; The spheres are the same size (have the same radii) so collisions between them are direct impacts.

5 a $\frac{9}{16}u, \frac{5}{16}u$

  b $\frac{3}{8}u$

6

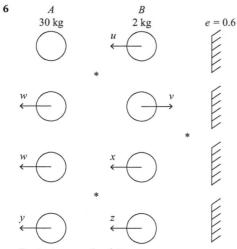

For 1st impact A and B

Momentum equation is

$2u + 0 = 30w - 2v$

$\qquad u = 15w - v \qquad\qquad$ (1)

Newton's equation is

$\dfrac{w+v}{u} = 0.6$

$\qquad 0.6u = w + v \qquad\qquad$ (2)

Solve (1): and (2): Add

$1.6u = 16w$

$\qquad w = \dfrac{u}{10}$

$\qquad v = \dfrac{u}{2}$

For impact of B on wall, Newton's equation

$x = 0.6v = 0.3u$

$x > w$ so B impacts on A

Momentum equation is

$30w + 2x = 30y + 2z$

$3u + 0.6u = 30y + 2z$

$\qquad 1.8u = 15y + z \qquad\qquad$ (3)

Newton's equation is

$\dfrac{y-z}{x-w} = 0.6$

$0.6\,(0.2u) = y - z \qquad\qquad$ (4)

Solve (3) and (4): Add

$1.8u + 0.12u = 16y$

$\qquad y = \dfrac{1.92u}{16} = 0.12u$

From (4)

$z = y - 0.12u = 0$

Hence, sphere B is at rest.

7 $1.33\,\mathrm{m\,s^{-1}}$, $5.15\,\mathrm{m\,s^{-1}}$ and $7.68\,\mathrm{m\,s^{-1}}$

No more collisions as P, Q and R are moving in the same direction with velocity $P <$ velocity $Q <$ velocity R

8 Using $v^2 = u^2 + 2\,as$ for the initial drop, $v^2 = 0 + 2 \times 10 \times 5 = 100$, so speed before first impact is $\sqrt{100} = 10\,\mathrm{m\,s^{-1}}$. Time between first and second impacts is 1 s. Total time is 3 s.

9 **a** $4\,\mathrm{m\,s^{-1}}$

b $2\,\mathrm{m\,s^{-1}}$

   X and Y together are moving in the same direction as Z but with less speed, therefore no more collisions.

Exercise 2.3A

1 **a** $40\,\mathrm{N\,s}$

   b $210\,\mathrm{N\,s}$

2 **a** $18\,\mathrm{N\,s}$

   b $45\,\mathrm{N\,s}$

3 **a** $960\,\mathrm{m\,s^{-1}}$

   b $232\,\mathrm{m\,s^{-1}}$

4 **a** $181\,\mathrm{N}$

   b $283\,\mathrm{N}$

5 0.12 N

6 a 32 N s, 8.4 m s^{-1}

 b 4 N s, 2.8 m s^{-1}

Exercise 2.3B

1 7.2 N s

2 1.2 m s^{-1}, 4.8 N s

3 3.6 N s, 1.5 N s

4 36 N s, 23 m s^{-1}

5 $42\dfrac{2}{3}$ N s, $20\dfrac{2}{3}$ m s^{-1}

6 240 N s, 5.6 m s^{-1}

7 a 48 N s

 b 2.4 m s^{-1}

 c 2.83 s

8 $\dfrac{Mm(u-v)}{m+M}$ N s

9 a 12 m s^{-1}, 36 N s

 b Initially, all the energy is the KE of Q (which equals $\dfrac{1}{2} \times 4 \times 21^2 = 882$ J). Finally, both P and Q have KE (which totals $\dfrac{1}{2} \times 7 \times 12^2 = 504$ J). But some KE will have been destroyed with the impulsive tension in the string when it tightened (amounting to $882 - 504 = 378$ J).

10 a 475 N s, 9.5 m s^{-1}

 b 2256.25 N

Exercise 2.4A

1 $\dfrac{5}{2}\mathbf{i} - 2\mathbf{j}$ m s^{-1}

2 $6\mathbf{i} + \mathbf{j}$ m s^{-1}

3 $\mathbf{v} = \mathbf{i} - 3\mathbf{j}$ m s^{-1}

4 $2\mathbf{i} + 2\dfrac{1}{4}\mathbf{j}$ m s^{-1} and $4\mathbf{i} + 4\dfrac{1}{2}\mathbf{j}$ m s^{-1}

5 $a = 18, b = 23$

6 a **A** $v_x = 4, v_y = 1{,}5$ **B** $v_x = 5, v_y = 0.5$ **C** $e = \dfrac{3}{4}, v_x = 3$ **D** $u_x = 4, e = \dfrac{1}{2}$

 b **A** 4.27 m s^{-1}, 20.6° **B** 5.02 m s^{-1}, 5.71° **C** 4.24 m s^{-1}, 45° **D** 4.12 m s^{-1}, 14.0°

7 $3\mathbf{i} + 2\mathbf{j}$ m s^{-1}, 86.8°

8 5.55 m s^{-1}, 25.6°

9 $2\mathbf{i} + 0.6\mathbf{j}$ m s^{-1}, 18 N s

10 $2.4\mathbf{i} + 3\mathbf{j}$ m s^{-1}, 38.4 N s

11 a $12\mathbf{i} + 9\mathbf{j}$ m s^{-1}

 b 175 J

Exercise 2.4B

1 $\dfrac{12\sqrt{2}}{5}$ (= 3.39) m s^{-1}

2 Impulse between A and B is along C_1C_2, so B begins to move in this direction along the x-axis, with velocity v. $3\sqrt{3}$ m s^{-1} and 6 m s^{-1}

3 $5\sqrt{3}$ m s^{-1} perpendicular to C_1C_2, 2.5 m s^{-1} along C_1C_2

4 5.99 m s^{-1}, 7.75 m s^{-1}, 12.1 J

5 a $\dfrac{1}{7}$

 b The balls and the surface they move on are all smooth. The balls have no spin.

6 2.37 J, 3.56 N s

7 3.00 m s^{-1} at 41.9° to C_1C_2, 3.87 m s^{-1} at 26.6° to C_1C_2, 6.07 J, 9.86 N s

8 a No impulse perpendicular to x-axis. Momentum conserved.

 b $\mathbf{i} + 3\mathbf{j}$ m s^{-1}, $2\mathbf{i}$ m s^{-1}, 4 N s

9 $4\mathbf{i} + 0.4\mathbf{j}$ m s^{-1}, $2.8\mathbf{j}$ m s^{-1}, 11.2 N s

10 a $\mathbf{i} + 2\mathbf{j}$ m s^{-1}, $2.5\mathbf{i} + \mathbf{j}$ m s^{-1}, 41.6°

 b 18 N s

11 $\dfrac{4\sqrt{3}}{3}\mathbf{i} + 4\mathbf{j}$ m s^{-1}, $\dfrac{7\sqrt{3}}{3}\mathbf{i} + \mathbf{j}$ m s^{-1}, $\dfrac{16\sqrt{3}}{3}$ N s

12 $5\sqrt{3}$ m s^{-1} perpendicular to line of centres, 2.5 m s^{-1} along line of centres

13 a $\frac{1}{3}u\sqrt{3}$ m s^{-1}, 180°

b

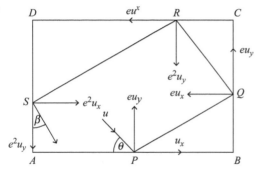

At P, components along and perpendicular to AB after impact are u_x and eu_y respectively.
Similarly at Q, components are eu_y along BC and $eu_x \perp^e BC$ after impact.
Similarly at R and S

At S, $\tan \beta = \dfrac{e^2 u_x}{e^2 u_y} = \dfrac{u_x}{u_y}$

At P, $\tan \theta = \dfrac{u_y}{u_x}$

So $\tan \beta = \cot \theta$ and $\theta + \beta = 90°$
and direction of u at P is same as final velocity at S
That is, these directions are parallel.

c The ball is made to spin as well as travel forward. An improved model would take into account the angular motion of the ball about its centre.

14 $\dfrac{1}{3}$

15 10.2 m s^{-1}, 33.7° below the horizontal

16 5.20 m s^{-1}, 5.05 m s^{-1}, 60.6 N s

17 a $\dfrac{u}{\sqrt{2}}$ m s^{-1}, $u\sqrt{\dfrac{3}{2}}$ m s^{-1}, $\dfrac{\sqrt{3}-1}{2} mu$ N s

b The model assumes that the spheres are smooth and that they have no spin.

18 a

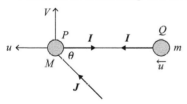

Let velocity of Q be u and components of velocity of P
be u and v
Impulse in rod $= I$
Impulse equation perpendicular to the rod is
$J \sin \theta = Mv$
Impulse equation along the rod for P and Q together is
$J \cos \theta = Mu + mu$

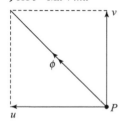

If angle of final velocity of P to rod $= \phi$

$\tan \phi = \dfrac{v}{u} = \dfrac{J \sin \theta}{M} \times \dfrac{M+m}{J \cos \theta}$

$\tan \phi = \dfrac{M+m}{M} \tan \theta$

b All contacts with the table are smooth; Impulse I acts parallel to the table so the impulsive thrust has no vertical component.

c A rod having a mass; A table which is not smooth.

19

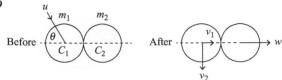

Let initial velocity of m_1 be u and final components be v_1 and v_2 as shown.
Along C_1C_2, momentum equation is
$$m_1 u \cos\theta = m_1 v_1 + m_2 w \qquad (1)$$
Newton's equation is
$$\frac{w - v_1}{u\cos\theta} = e$$
$$w - v_1 = e u \cos\theta \qquad (2)$$
$\perp^e C_1C_2$ no impulse, so no change in velocities $v_2 = u \sin\theta$
Eliminate w from (1) and (2)
$$m_1 u \cos\theta = m_1 v_1 + m_2(v_1 + eu\cos\theta)$$
$$u\cos\theta(m_1 - em_2) = (m_1 + m_2)v_1$$
Now $v_1 = 0$ if $m_1 = em_2$
in which case the final velocities are v_2 and w which are $\perp^e$ to each other.

20 $\dfrac{1}{2}mu^2\left(1 - e^2\right)$

For direct impact,

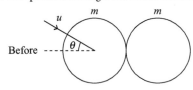

Newton's equation is
$$\frac{2v}{2u} = e, \quad v = eu$$

Loss of KE $= 2 \times \dfrac{1}{2}mu^2 - 2 \times \dfrac{1}{2}m(eu)^2$
$$= mu^2(1 - e^2)$$

So loss of KE when oblique $= \dfrac{1}{2} \times$ loss of KE when direct impact

21 a $\dfrac{mu\sqrt{3}}{2}, \dfrac{u}{2}$

   b $\dfrac{mu\sqrt{3}}{4}, \dfrac{\sqrt{7}}{4}u$

22 Let speed u be at angle θ to line of centres.

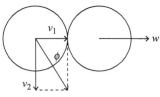

Let final velocities be w along the line of centres and $v_1\mathbf{i} + v_2\mathbf{j}$ at an angle ϕ to line of centres.
No impulse perpendicular to the line of centres
so $v_2 = u\sin\theta$
Momentum equation along the line of centres is
$$mu\cos\theta = mw + mv_1$$
so $w + v_1 = u\cos\theta \qquad (1)$
Newton's equation perpendicular to the line of centres is
$$\frac{w - v_1}{u\sin\theta} = \frac{1}{3} \qquad (2)$$
$$w - v_1 = \frac{1}{3}u\cos\theta$$
Subtract (2) from (1)
$$2v_1 = \frac{2}{3}u\cos\theta$$

$$v_1 = \frac{1}{3}u\cos\theta$$

$$\tan\phi = \frac{v_2}{v_1} = \frac{u\sin\theta}{\frac{1}{3}u\cos\theta} = 3\tan\theta$$

23

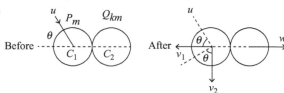

Along C_1C_2, momentum equation is

$$mu\cos\theta + 0 = -mv_1 + kmw$$

$$u\cos\theta = kw - v_1 \qquad (1)$$

Newton's equation is

$$\frac{w + v_1}{u\cos\theta} = e$$

$$eu\cos\theta = w + v_1 \qquad (2)$$

Add (1) and (2):

$$(1+k)w = u\cos\theta \cdot (1+e)$$

From (2)

$$v_1 = eu\cos\theta - w$$

$$= eu\cos\theta - \frac{u\cos(1+e)}{1+k}$$

$$= \frac{(1+k)eu\cos\theta - (1+e)u\cos\theta}{1+k}$$

$$= \frac{ek-1}{k+1}\cdot u\cos\theta$$

As P is deflected through 90°, v_1 in diagram is positive.

$v_1 > 0$ so $ek - 1 > 0$, $e > \dfrac{1}{k}$

$\perp^r C_1C_2$ no impulse, so no change in velocity, $v_2 = u\sin\theta$

After impact, $\tan\theta = \dfrac{v_1}{v_2}$

$$= \frac{ek-1}{k+1}\cdot\frac{u\cos\theta}{u\sin\theta}$$

giving $\tan^2\theta = \dfrac{ek-1}{k+1}$

so $e = \dfrac{\tan^2\theta\,(k+1)+1}{k}$

As $\dfrac{1}{k} < e \le 1$

so $1 < \tan^2\theta(k+1) + 1 \le k$

$0 < \tan^2\theta \times (k+1) \le k - 1$

and $\tan^2\theta \le \dfrac{k-1}{k+1}$

Review exercise 2

1 $-6\ \text{m s}^{-1}$

2 $(-)9\dfrac{1}{4}\ \text{m s}^{-1}$, $3\dfrac{3}{4}\ \text{m s}^{-1}$

3 $\dfrac{1}{3}$, $11\dfrac{2}{9}\ \text{m}$

4 $2\ \text{m s}^{-1}$, $10\,000\ \text{N s}$

5 $10\dfrac{1}{2}\ \text{kg}$, $\dfrac{4}{5}$, $42\ \text{N s}$

6 $18750\ \text{N}$

7 $6\ \text{N s}$, $12\ \text{m s}^{-1}$

8 After the first collision, B has velocity $8\ \text{m s}^{-1}$ to the left. Since A is at rest, there will be a collision with A. The velocities after the second collision are $8\ \text{m s}^{-1}$ to the left, $4\ \text{m s}^{-1}$ to the left, $1\ \text{m s}^{-1}$ to the right.
As $8 > 4 > -1$ there are no more collisions.

9 $\sqrt{10}$ m s⁻¹, −18.4°

10 10 kg, $\begin{pmatrix} \dfrac{4}{3} \\[2mm] \dfrac{4}{3} \end{pmatrix}$ m s⁻¹

11 $5\mathbf{i} + \mathbf{j}$ m s⁻¹, 12 N s

12 $4\mathbf{i}$ m s⁻¹, $4\mathbf{i} + 5\mathbf{j}$ m s⁻¹, 8 N s

13 13.0°, 14 N s

14 8.89 m s⁻¹ at 41.9° to C_1C_2, 11.6 m s⁻¹ at 26.6° to C_1C_2, 54.6 J

15

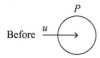

Along line of centres C_1C_2, momentum equation is
$$mu + 0 = mv_1 + mw_1$$
$$u = v_1 + w_1 \qquad\qquad (1)$$
Newton's equation is
$$\frac{w_1 - v_1}{u} = e$$
$$w_1 - v_1 = eu \qquad\qquad (2)$$
Add (1) and (2)
$$2w_1 = u(1 + e)$$
Subtract (2) from (1)
$$2v_1 = u(1 - e)$$
Perpendicular to C_1C_2, no impulse, so no change in velocity
$$\Rightarrow v_2 = 0 \quad \text{and} \quad w_2 = u$$
Angle turned through by Q
$$\theta = \tan^{-1}\left(\frac{w_1}{w_2}\right)$$
$$= \tan^{-1}\left(\frac{u(1+e)}{2} \times \frac{1}{u}\right)$$
$$= \tan^{-1}\left(\frac{1+e}{2}\right)$$

16 $\dfrac{I}{2m}$ at 45° (to the horizontal), $\dfrac{I}{2\sqrt{2}m}$, 0, $\dfrac{I}{2\sqrt{2}m}$

Assuming light, inextensible strings.

Assessment 2

1 **a** 42 N s

    b 120 N

2 **a** 4 s

    b The particles are smooth and have no spin.

3 **a** $(4.4\mathbf{i} - 0.4\mathbf{j})$ m s⁻¹

    b There is no rebound; The ball is reduced to rest on striking the wall; It loses all its KE.

4 4.5 s

5 11.4 m s⁻¹ at 18.1° to the cushion

6 **a** 3 m s⁻¹

    b 4.76 m s⁻¹

7 **a** $\dfrac{10v}{7}$

    b $\dfrac{2v}{7}$

8 **a** $\dfrac{14}{3}$ m s⁻¹

    b $50 - 2m = 30 + mv$ gives $v = \dfrac{20 - 2m}{m}$

      B must travel at least as fast as A

      so $\dfrac{20 - 2m}{m} \geq 3 \;\Rightarrow\; m \leq 4$

9 a $(4\mathbf{i} - 1.8\mathbf{j})\,\mathrm{m\,s^{-1}}$, $(-\mathbf{i} + 2.2\mathbf{j})\,\mathrm{m\,s^{-1}}$
b $14.4\mathbf{j}\,\mathrm{N\,s}$
c $-61.1°$
10 a $15\,\mathrm{m\,s^{-1}}$
b $6000\,\mathrm{N}$
c 5882
d The post intially travels faster than before (the hammer has less momentum, so the post must have greater momentum). The post moving on its own will be brought to rest more rapidly by the constant resistance force.
11 $(2.5\mathbf{i} + 0.5\mathbf{j})\,\mathrm{m\,s^{-1}}$
The ball strikes the wall without any spin; Its final velocity is immediately after striking the wall and not some time later.
12 a i $4\,\mathrm{m\,s^{-1}}$, $9\,\mathrm{m\,s^{-1}}$ **ii** $6\,\mathrm{m\,s^{-1}}$
b i No momentum is lost, so final momentum $5v$ equals initial momentum 30 whatever the value of e
ii $e = 0$ means the particles would not separate after collision, so the string would never become taut. The particles would move together at $6\,\mathrm{m\,s^{-1}}$
13 a $7.5\,\mathrm{m\,s^{-1}}$, $10\,\mathrm{m\,s^{-1}}$
b 0, $7.5\,\mathrm{m\,s^{-1}}$
c Following the third collision, A moves with a speed of $2.8125\,\mathrm{m\,s^{-1}}$ to the left, B is at rest. So no more collisions.
14 a $2.2\,\mathrm{m\,s^{-1}}$, $3.2\,\mathrm{m\,s^{-1}}$
b $\dfrac{7.4 + 2.2e}{3}\,\mathrm{m\,s^{-1}}$
c If there are no more collisions, B must be moving faster than A.
$$\dfrac{7.4 + 2.2e}{3} \geq 2.2 \quad \Rightarrow \quad e \leq \dfrac{4}{11}$$
15 a i The cushion is smooth, so no force and hence no change of momentum in the **i**-direction.
ii $4.9\,\mathrm{m\,s^{-1}}$ **iii** $98.9°$
b 2380
c As the x-component of $\mathbf{v}$ is $5\,\mathrm{m\,s^{-1}}$, in a contact time of $0.1\,\mathrm{s}$ the ball would slide $0.5\,\mathrm{m}$ along the cushion, which is unrealistic.
16 a $\dfrac{2 - \sqrt{3}}{\sqrt{3}} = 0.155$
b $4.96\,\mathrm{ms^{-1}}$ at $53.8°$ to line of centres
17 a $4\,\mathrm{m\,s^{-1}}$
b $6\,\mathrm{N\,s}$
c The total momentum of A and B is now $1 \times 4 + 2 \times 1 = 6\,\mathrm{N\,s}$. This will be unchanged by the collision between A and B and the string again becoming taut.
Their common velocity is then v, where $3v = 6$, $v = 2\,\mathrm{m\,s^{-1}}$
They are moving slower than C, so there are no more collisions.
18 $\dfrac{u}{4}$, $\dfrac{u}{2}$
19 4.25, 0.3125
20 a $6\,\mathrm{kg}$
b $1.8\,\mathrm{m\,s^{-1}}$
21 $7.2\,\mathrm{N\,s}$
22 a $2\,\mathrm{m\,s^{-1}}$
b Momentum in the vertical direction would not be conserved. The gun would recoil more slowly, as the magnitude of its momentum would equal the horizontal component of the shell's momentum.
23 a B rebounds from wall with speed $0.4 \times 10 = 4\,\mathrm{m\,s^{-1}}$
When A and B collide the momentum equation is
$4 - 8 = 4v_A + 2v_B$ [1]
The restitution equation is $0.2 \times 5 = 1 = v_B - v_A$ [2]
Solving [1] and [2]
$v_A = -1$, $v_B = 0$
b $1\,\mathrm{m\,s^{-1}}$
24 $-8\mathbf{j}\,\mathrm{m\,s^{-1}}$
25 a $5.6\,\mathrm{m\,s^{-1}}$
b 0.2
26 $7.38\,\mathrm{m\,s^{-1}}$, $4.62\,\mathrm{m\,s^{-1}}$
27 $3.00\,\mathrm{ms^{-1}}$ at $41.9°$ to line of centres, $3.87\,\mathrm{ms^{-1}}$ at $26.6°$ to line of centres
28 $(10\mathbf{i} + 20\mathbf{j})\,\mathrm{N\,s}$
$|I| = \sqrt{10^2 + 20^2} = \sqrt{500} = 10\sqrt{5}\,\mathrm{Ns}$
29 a $68\,\mathrm{N\,s}$
b $10.5\,\mathrm{m\,s^{-1}}$
30 $2.7\,\mathrm{J}$
31 $49.5\,\mathrm{J}$

Index

2D motion 46–54

A
acceleration 9, 28
air resistance 3
approach speed 34–5
average power 3, 5

B
calculus 40–1
coefficient of restitution (*e*) 34–7, 47, 51
collisions 27–63
    2D motion 46–54
    conservation of momentum 27–33,
        36–7, 46–7, 50–1
    $Ft = mv - mu$ 28–9, 40–1
    impulses 28, 40–54
compressions 12
conservation of mechanical energy 3, 15–16
conservation of momentum 27–33, 36–7,
    46–7, 50–1
constant acceleration formula 9
constant forces 2–3, 6, 8, 28, 40–1
cosine (cos), work done formula 8

D
differentiation 40–1
direct collisions 29, 50
direction of force 2–3, 6
distance 2–3, 6
dragging an object 2, 4, 6

E
e (coefficient of restitution) 34–7, 47, 51
elasticity 36
    elastic limit 12
    EPE 13, 15–16
    inelasticity 34, 48
    modulus of 12–13
    perfect 34
    strings 12–13, 15–16
elastic potential energy (EPE) 13, 15–16
energy 1–26
    forces 1–26
    Hooke's law 12–18
    mechanical 3–4, 15–16
    power 2–11
    work 2–11
EPE *see* elastic potential energy
equilibrium 13, 15
extensions 12–13, 15
external impulses 48, 50

F
forces 1–26
    energy 1–26
    Hooke's law 12–18
    power 2–11
    work 2–11
friction 2, 4, 15

G
GPE *see* gravitational potential energy
graphs 12, 44
gravitational potential energy (GPE) 3–5, 8,
    15–16

H
Hooke's law 12–18

I
impulses 28, 40–54
    2D motion 46–54
    $Ft = mv - mu$ 28, 40–1
    impulse equations 40–1, 44, 48–9
    units 28
inelasticity 34, 48
integration 12, 40–1, 44
internal impulses 48, 50

J
joules 2–3

K
kinetic energy (KE)
    collisions 36–7
    energy 3–5, 8
    Hooke's law 13, 15–16

M
mechanical energy 3–4, 15–16
modulus of elasticity 12–13
momentum 27–63
    2D motion 46–54
    conservation of 27–33, 36–7, 46–7, 50–1
    $Ft = mv - mu$ 28–9, 40–1
    impulses 40–54
    momentum equations 31, 34–7, 43, 47,
        50–1
    units of 28
motion equations 3, 9

N
natural length of string 12–13, 15
newtons 12
newton seconds (Ns) 28
Newton's laws
    2nd law 3, 9, 28, 40
    3rd law 28, 40–1
    law of restitution 34–7, 47, 50–1
Ns *see* newton seconds

O
oblique impacts 50

P
parallel motion 47
PE *see* potential energy
perfect elasticity 34
perpendicular motion 47
potential energy (PE) 3–5, 8, 13, 15–16

power 2–11
    average power 3, 5
    energy 2–11
    forces 2–11
    units of 3
principle of conservation of (linear)
    momentum 29
principle of conservation of mechanical
    energy 3, 15
Pythagoras's theorem 46

R
resistance 2, 4, 8
restitution 34–7, 47, 50–1

S
separation speed 34–5
simultaneous equations 13, 35, 37, 51
sine (sin) ratio 8
speed 3–5, 8, 41, 46
    of approach 34–5
    of separation 34–5
springs 12–13, 15
stiffness 12
strings 12–13, 15–16, 43, 48

T
T see tension
tangent (tan) ratio 46
tension (*T*) 2, 4
    Hooke's law 12–13, 15
    impulses 43, 48
tractive forces 2, 4–6
trigonometric ratios 8, 46

V
variable forces 2–3, 44
vectors 28, 46, 49, 50
velocity
    applying force to mass 3
    collisions 28–9, 31, 34, 36–7
    conservation of momentum 28–9, 31
    Hooke's law 15–16
    impulses 41, 43, 44, 46–8, 50–1

W
watts (W) 3
work
    energy 2–11
    forces 2–11
    $F s \cos \theta$ 6
    Hooke's law 12–13
    units of 2
    work done 2–6, 8, 13